Lessons learned from Long-term Soil Fertility Management Experiments in Africa

Andre Bationo • Boaz Waswa • Job Kihara
Ivan Adolwa • Bernard Vanlauwe • Koala Saidou
Editors

Lessons learned from Long-term Soil Fertility Management Experiments in Africa

Springer

Editors
Andre Bationo
Alliance for a Green Revolution
in Africa (AGRA)
Resource Mobilization Office, CSIR Office
Complex #6 Agostino Neto Road Airport
Residential Area PMB KIA 114 Airport
Accra, Ghana

Boaz Waswa
Tropical Soil Biology & Fertility
Institute of International Centre
for Tropical Agriculture (CIAT-TSBF)
ICIPE Duduville Campus off
Kasarani Road
Nairobi, Kenya

Job Kihara
Tropical Soil Biology & Fertility
Institute of International Centre
for Tropical Agriculture (CIAT-TSBF)
ICIPE Duduville Campus off
Kasarani Road
Nairobi, Kenya

Ivan Adolwa
Tropical Soil Biology & Fertility
Institute of International Centre
for Tropical Agriculture (CIAT-TSBF)
ICIPE Duduville Campus off
Kasarani Road
Nairobi, Kenya

Bernard Vanlauwe
Tropical Soil Biology & Fertility
Institute of International Centre
for Tropical Agriculture (CIAT-TSBF)
ICIPE Duduville Campus off
Kasarani Road
Nairobi, Kenya

Koala Saidou
Tropical Soil Biology & Fertility
Institute of International Centre
for Tropical Agriculture (CIAT-TSBF)
ICIPE Duduville Campus off
Kasarani Road
Nairobi, Kenya

ISBN 978-94-007-2937-7 ISBN 978-94-007-2938-4 (eBook)
DOI 10.1007/978-94-007-2938-4
Springer Dordrecht Heidelberg New York London

Library of Congress Control Number: 2012934458

Printed on acid-free paper

Springer is part of Springer Science+Business Media (www.springer.com)

Preface

Long-term experiments (LTE) play an important role in understanding the effects of land management on soil fertility, sustainability of yield or wider environmental issues. It may take decades to reveal an ongoing negative trend of depletion of a soil's productivity. Similarly, it can take an equally long time to rehabilitate and restore productivity of land. Monitoring such changes can only be achieved if long term experiments that are systematically managed without compromising their scientific integrity existed.

The Classical LTEs at Rothamsted Research in the UK are among the oldest existing agronomic experiments in the world dating back to 1843. Other old LTE include the Morrow Plots, located at the University of Illinois in the US, which were established in 1876. Countries such as Sweden have had LTE since 1930s and 1950s. Only a handful LTE have been established and maintained to date in Sub-Saharan Africa (SSA). Where such LTE exist, data from such experiments are scattered, difficult to access and have remained unpublished. The experiments continue to suffer from inconsistency in management as well as poor funding. Despite these limitations, there exist some potential LTE from which lessons on changes in soil properties and crop productivity can be assessed. The Samaru, Zaria experiment in Nigeria, established in 1950 and the Saria experiment in Burkina Faso, established in 1960 are among the oldest soil fertility experiment in Africa. Other experiments are the Long Term Experiment at the National Agriculture Research Laboratories (NARL), Kabete, Kenya established in 1976. Most other experiments are of intermediate age ranging from 5-10 years.

The LTE in Africa have been established in a wide range of agroecological zones (AEZ) ranging from the tropical humid to the arid and semi arid regions. These AEZ are characterized by a wide range of soil types. Most experiments have targeted the main cereal (maize, millet, sorghum, rice) and legume (beans, cowpea, groundnuts, pigeon pea) crops in the respective regions. The experiments cover a wide range of integrated soil fertility management (ISFM) technologies such as efficient use of inorganic fertilizer, combining organic and inorganic nutrient sources, integration of legumes in cropping systems, use of manure and crop residues, use of rock phosphate and other soil amendments, soil and water conservation among others. Beyond responding to technical and biophysical components of the experiments, effort has been made to understand the socio-economic conditions under which the promising technologies can be widely adopted.

This book gives an evaluation of experiences from selected LTE in East South and West Africa. The first chapter presents an overview and justification for LTE using lessons from selected experiments. The second, sixth, eighth, ninth and tenth chapters give an in-depth account of crop management practices and soil fertility interventions in long-term trials within specific agro-ecological zones in West Africa. The rest of the chapters (chapter three, four, five and seven) address crop management, tillage practices and, organic and inorganic fertilizer applications in the context of long-term experiments in specific agro-ecological zones in East Africa. It is our hope that the lessons from these experiments will elicit revived interest in LTE in Africa not only for their scientific wealth but for promoting sustainable land use management practices that will cushion farmers against emerging global environmental changes.

His Honorable Laurent Sedogo
Minister of Agriculture and Hydraullique, Water and Fisheries, Burkina Faso
&
His Honorable Gnissa Konate
Minister of Scientific Research and Innovation, Burkina Faso

Contents

Chapter 1
Overview of Long Term Experiments in Africa

A. Bationo, B. Waswa, A. Abdou, B.V. Bado, M. Bonzi, E. Iwuafor, C. Kibunja, J. Kihara, M. Mucheru, D. Mugendi, J. Mugwe, C. Mwale, J. Okeyo, A. Olle, K. Roing, and M. Sedogo

Abstract The prevailing low food production in sub-Saharan Africa is an issue of great concern especially since Africa south of the Sahara is the only remaining region of the world where per capita food production has remained stagnant. This chapter reviews long-term experiments in Africa in the context of shifting paradigms related to tropical soil fertility management from first external input paradigm right through to the current Integrated Soil Fertility Management (ISFM) approach, which is a culmination of the participatory methods developed along the paradigm shift. Long term experiments (LTE) are an important source of evidence for soil fertility decline and provide crucial datasets for the development of sustainable management practices for tropical land-use systems and the amelioration of global climatic and environmental change impacts. A survey was undertaken to identify

---

A. Bationo (✉)
Alliance for a Green Revolution in Africa (AGRA), Accra, Ghana
e-mail: abationo@agra-alliance.org

B. Waswa
Tropical Soil Biology & Fertility Institute of International Centre for Tropical Agriculture (CIAT-TSBF), ICIPE Duduville Campus off Kasarani Road, P.O. Box 823-00621, Nairobi, Kenya

University of Bonn, Bonn, Germany

A. Abdou
Tropical Soil Biology and Fertility (TSBF), International Center for Research in the Semi-Arid Tropics (ICRISAT), Niamey, Niger

B.V. Bado
AfricaRice, BP 96, Saint Louis, Senegal

M. Bonzi • M. Sedogo
INERA, 04 BP, 8645 Ouagadougou 04, Burkina Faso

E. Iwuafor
Ahmadu Bello University, Zaria, Nigeria

A. Bationo et al. (eds.), *Lessons learned from Long-term Soil Fertility Management Experiments in Africa*, DOI 10.1007/978-94-007-2938-4_1,

some ongoing long-term trials distributed across east, south and western Africa and in different agro-ecological zones. A long-term Experiment was defined as that extending over a period of over 5 years and more. Results from these trials are discussed in detail. Inference has been drawn from these findings and includes the following key findings: a) All long term trials showed yield decline, often with a relatively rapid fall to a low level equilibrium; b) At all sites, there were positive yield responses to one or more nutrients added as mineral fertilizers, which were consistent for the duration of the experiments highlighting the effectiveness of mineral fertilizers in increasing yield in arable farming systems in Africa; c) Soil organic matter (SOM) also declines significantly when land is cultivated; d) Prolonged treatments using only inputs of organic matter also showed yield declines, although the positive impact were sustained longer than for inorganic fertilizers alone in most cases; e) Rotational treatments, including sequences with legume crops and fallow periods had lower yield declines than monocultures; f) The best results invariably were those treatments that combined inorganic and organic inputs.

Keywords Long-term experiments • Integrated soil fertility management • Soil organic matter • Inorganic fertilizers • Organic inputs

Introduction

Africa south of the Sahara is the only remaining region of the world where per capita food production has remained stagnant over the past 40 years (Sanchez 2002). About 180 million Africans – up 100% since 1970 – do not have access to sufficient food to lead healthy and productive lives, making them more susceptible to the ravages of malaria, HIV-AIDS, and tuberculosis. Absolute poverty-characterized by

C. Kibunja
Kenya Agricultural Research Institute National Agricultural Research Laboratories (KARI-NARL), Nairobi, Kenya

J. Kihara • K. Roing
Tropical Soil Biology and Fertility Institute of International Centre for Tropical Agriculture (CIAT-TSBF), ICIPE Duduville Campus off Kasarani Road, P.O. Box 823-00621, Nairobi, Kenya

M. Mucheru • J. Mugwe
Department of Environmental Resource Conservation, Kenyatta University, Nairobi, Kenya

D. Mugendi
Kenyatta University, Nairobi, Kenya

C. Mwale
Zambia Agricultural Research Institute, Chilanga, Zambia

J. Okeyo
University of Wyoming, Wyoming, USA

A. Olle
Department of Soil Sciences, SLU, Uppsala, Sweden

incomes of less than U.S. $1 per person per day is coupled with an increasingly damaged natural resource base (Pinstrup-Andersen et al. 2000). Low food production in the continents is as a result of the breakdown of traditional practices and the low priority given by governments to the rural sector (Sanchez et al. 1997).

Over decades, small-scale farmers have removed large quantities of nutrients from their soils without using sufficient quantities of manure or fertilizer to replenish the soil. This has resulted in a very high average annual depletion rate – 22 kg of nitrogen (N), 2.5 kg of phosphorus (P), and 15 kg of potassium (K) per hectare of cultivated land over the last 30 years in 37 African countries – an annual loss equivalent to U.S. $4 billion in fertilizer (Sanchez et al. 1997; Lynam et al. 1998). The potential of genetically improved crops cannot be realized when soils are depleted of plant nutrients. A recent study shows that while the rates of adoption of improved crop varieties have been similar in Asia, Latin America, the Middle East, and Sub-Saharan Africa during the last 38 years, such varieties are responsible for 66–88% of the crop yield increases in the first three regions, but only 28% in Africa (CGIAR 2001). As a result there is need to invest more in soil fertility management if Africa is to benefit from the benefits improved crop varieties.

Paradigm Shifts Related to Tropical Soil Fertility Management

Over the years, the paradigms underlying soil fertility management research and development efforts have undergone substantial change because of experiences gained with a specific approaches and changes in the overall social, economic, and political environment the various stakeholders are facing.

During the 1960s and 1970s, an external input paradigm was driving the research and development agenda. The appropriate use of external inputs, be it fertilizers, lime, or irrigation water, was believed to be able to alleviate any constraint to crop production. Following this paradigm together with the use of improved cereal germplasm, the 'Green Revolution' boosted agricultural production in Asia and Latin America in ways not seen before. Organic resources were considered less essential. However, application of the 'Green Revolution' strategy resulted only in minor achievements in Sub-Saharan Africa (SSA). Concerns of environmental degradation resulting from massive applications of fertilizers and pesticides in Asia and Latin-America between the mid-1980s and early-1990s (Theng 1991) and the abolition of the fertilizer subsidies in SSA (Smaling 1993), imposed by structural adjustment programs led to a renewed interest in organic resources in the early 1980s. The balance shifted from mineral inputs only to low mineral input sustainable agriculture (LISA) where organic resources were believed to enable sustainable agricultural production. The potential of LISA was hindered by several constraints both at the technical (e.g., lack of sufficient organic resources) and the socio-economic level (e.g., labour intensive technologies).

In this context, Sanchez (1994) revised his earlier statement by formulating the Second Paradigm for tropical soil fertility research that recognized the need for both mineral and organic inputs to sustain crop production, and emphasized the need for all inputs to be used efficiently. The need for both organic and mineral inputs was advocated because (i) both resources fulfil different functions to maintain plant growth, (ii) under most small-scale farming conditions, neither of them is available or affordable in sufficient quantities to be applied alone, and (iii) several hypotheses could be formulated leading to added benefits when applying both inputs in combination. The second paradigm also highlighted the need for improved germplasm, as in earlier days, more emphasis was put on the nutrient supply side without worrying too much about the demand for these nutrients.

From the mid-1980s to the mid-1990s the shift in paradigm towards the combined use of organic and mineral inputs was accompanied by a shift in approaches towards involvement of the various stakeholders in the research and development process, mainly driving by the 'participatory' movement. One of the important lessons learnt was that the farmers' decision making process was not merely driven by the soil and climate but by a whole set of factors cutting across the biophysical, socio-economic, and political domain. This understanding led to evolution of the Sustainable Livelihoods Approach (DFID 2000), the Integrated Natural Resource Management (INRM) research approach and eventually the Integrated Soil Fertility Management (ISFM) paradigm. ISFM forms an integral part of the INRM research approach with a focus on appropriate management of the soil resource. Although technically ISFM adopts the Second Paradigm, it recognizes the important role of social, cultural, and economic processes regulating soil fertility management strategies. ISFM is also broader than Integrated Nutrient Management (INM) as it recognizes the need of an appropriate physical and chemical environment for plants to grow optimally, besides a sufficient and timely supply of available nutrients.

Justification for Long-Term Trials

Long term experiments (LTE) and monitoring are an important source of evidence for soil fertility decline. The studies provide the most convincing set of data as they highlight trends and dynamics rather than the static snapshots of most other measures. LTE serve as living laboratories providing opportunities for experimentation in which the effects of manipulation may be separated from other variables (Southwood 1994). This is clearly essential in understanding processes of soil fertility change. Long-term experiments can be viewed as living laboratories that provide opportunities for experimentation in which the effects of manipulation may be separated from other variables. Once a system is understood in this amount of detail, it is possible to manipulate almost any part and to be confident that one is able to assess the corresponding impacts.

An entirely different product of long term experiments has been the development of many of the statistical techniques used today (Southwood 1994). In the last few decades, Rothamsted has developed efficient statistical methods to analyse large data sets, methods that owe their origins to the seedbed of long-term experiments, but which are now used worldwide. With these statistical tools and this background, models can be developed which permit the inclusion of time into decision-making processed.

The increasing importance accorded to the development of sustainable management practices for tropical land-use systems and the apprehension of the potential impact of global climatic and environmental change has raised new interest in the datasets from these experiments as well as the possibilities for new initiatives in long-term monitoring and experimentation (Swift et al. 1994).

Long Term Experiments in Africa

There have been a substantial number of long-term experiments in Africa addressing a wide range of purposes. There is however no comprehensive inventory of these experiments. A number of these trials are still extant and actively researched, some have been judged to have reached the end of their useful existence and yet others have been discontinued or diminished in intensity because of lack of resources. The existence of these experiments provides opportunities to investigate the effects of different nutrient management practices on soil fertility and crop yields. Several reviews have been done on long term trials across Africa (Swift et al. 1994; Bekunda et al. 1997; Greenland 1994; Pieri 1995). Ever since more trials have been established and some key lessons from the old and new trials will be discussed in the next section.

A majority of the ongoing or terminated long-term soil fertility management trials in Africa were designed to determine the effects of inorganic fertilizers and organic inputs on crop yields and soil properties. Effects of rotations have also been investigated. Although yields were measured in all the experiments, climatic and soil variables were documented in only a few trials. The highest frequency of soil analysis was biannual, with most trials reporting soil analysis at only the beginning and end of the trial. No evidence of other measurement factors outside the treatments variable which would influence yields, such as pests, diseases incidences and economic parameters were measured.

This paper evaluates the experiences from a selection of ongoing long term experiments in east, south and western Africa.

Methodology

A survey was undertaken to identify some ongoing long-term trials in Africa. The trials are distributed across the content- east, south and western- and in different agroecological zones (Fig. 1.1). For the purpose of this review, we define long-term experiments as those extending over a period of over 5 years and more. Table 1.1

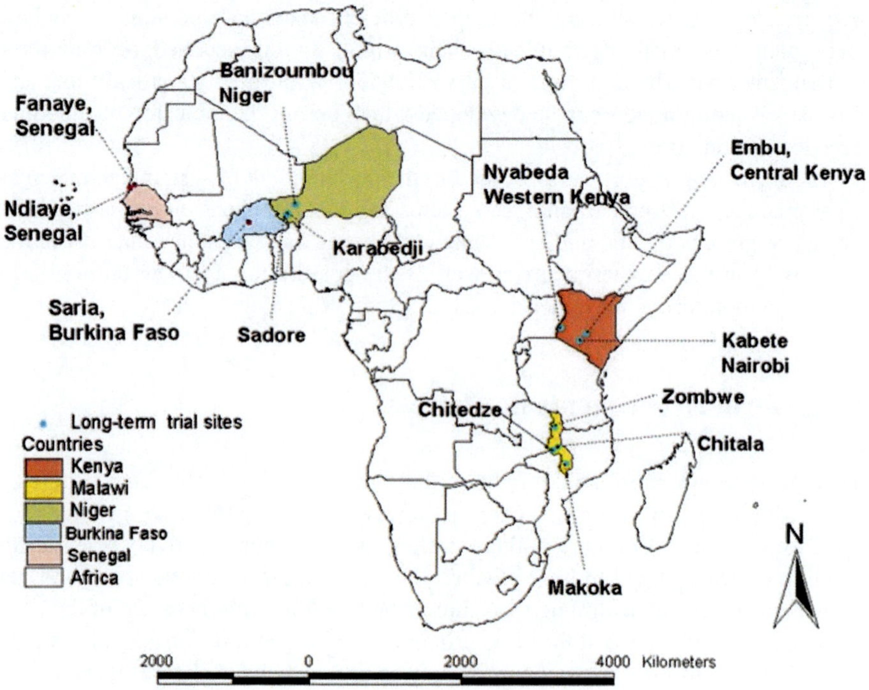

Fig. 1.1 Map showing location of selected long term trials in Africa

gives the general site information of some of these trials. The trials have varying lifespans and were established to achieve different objectives. Results from these trials are discussed in the subsequent section.

Long-Term Trials in West Africa

Long Term Trials in Saria, Burkina Faso

The long term trial in Saria Burkina Faso was started in 1960 with the purpose of evaluating the fertility of soils under different management systems. Sorghum and peanuts were the main test crops until 1987 when the legume was changed from peanuts to cowpea (Fig. 1.2).

An assessment of the long-term (over 20 years) effect of organic and inorganic fertilization on soil organic matter (SOM) fractions and sorghum performance resulted into different SOM concentrations (Mando 2006; Mando et al. 2005). Twenty years of continuous cultivation without external inputs (the control treatment) depleted SOM levels to below 50% of those under fallow. Sole urea application further depleted SOM status, presumably because of alleviation of nitrogen

Table 1.1 General information about the long-term trials in Africa

Experiment	Site/location	Year of est.	No. of cropping seasons	Longitude	Latitude	AEZ	Av Rainfall	Temp.		Initial soil properties				
								Max	Min	Soil Type	C (%)	N (mg/kg)	P (mg/kg)	pH
Long term trials in Saria	Saria, Burkina Faso	1960		2°9′W	12°16′N	Southern Sahelien	800	45	30					
Long-term crop residue management trial	Sadore, Niger	1982	25	2°17′E	13°14′N	Sahelian	560	30–41	17–27	Sandy	0.12	3	2	4.3
CR: Long-term effect of manure, crop residue and fertilizers	Sadore, Niger	1993	14	2°17′E	13°14′N	Sahelian	560	30–41	17–27	Sandy	0.12	0.3	2	4.3
Kouaré cowpea trial	Kouaré, Burkina Faso	1995	5	0°19′W	1159′N	Southern Sahelien					0.5	427 kg N		5.5
Effect of degraded and non degraded soils on millet production	Karabedji, Niger	1999	08	2°21′E	13°18′N	Sahelian	450	30–41	17–27	Sandy	0.16	4	1.9	4.2
N&P: fertilizer N and P equivalency of manure on millet	Banizoumbou, Niger	2001	06	3°19′E	14°21′N	Sahelian	360	30–41	17–27	Sandy	0.12	5	1.5	4.4
OPSCAR: operational scale research	Sadore, Niger													
Long-term fertility experiment trial	Ndiaye/Senegal	1991	32	16°14′W	16°14′N		180	45	10	Gleysol	1	0.08%	4	6.5

(continued)

Table 1.1 (continued)

Experiment	Site/location	Year of est.	No. of cropping seasons	Longitude	Latitude	AEZ	Av Rainfall	Temp.		Initial soil properties				
								Max	Min	Soil Type	C (%)	N (mg/kg)	P (mg/kg)	pH
Long-term fertility experiment trial	Fanaye/Senegal	1991	32	15°46′W	16°33′N		230	45	7	Vertisol	0.7	0.05%	5	6.8
Kabete long term trial	National Agricultural Research Laboratories (NARL), Kabete, Kenya Kenya	1976		36°46′ E	01°15′ S		950							
HI Embu trial	Embu Regional Research Centre, Embu, Kenya	1992	22	37°30′ E	0°30′S		1,200–1,500	25	14					
N1 Kabete trial	National Agricultural Research Laboratories (NARL), Kabete, Kenya Kenya	1999	15	36°46′ E	01°15′ S		950	23.8	12.6	Loam	1.6	1.35	27	5.4
Kirege trial	Chuka, Eastern Kenya	2000	12	36°46′ E	01°15′S		1,200–1,400	20						
INM2 trial	Maseno, Western Kenya	2001												
CT1 trial	Nyabeda, Western Kenya			34°34′E	0° 06′N									
Chitedze trial	Chitedze, Malawi	1995/1996	12	33°38′E	13° 59′S	Subtropical mid altitudes	892	24	16	Sand clay loam	2.39	0.16		5.4
Chitala trial	Chitala, Malawi	1995/1996	12	34°15′E	13°39′S	Subtropical lowlands	800	28	16	Sand clay loam	1.07	0.07		5.6
Makoka trial	Makoka, Malawi	1998/1999	9	35° 11′E	15°32′S	Subtropical lowlands	1,044	34.4	8.5	Sandy loam	0.30	0.02		5.3
Zombwe trial	Zombwe, Malawi	1998/1999	9	33°49′E	11°19′S	Subtropical mid altitudes	1,035	27	17	Sand clay loam	0.68	0.05		5.3

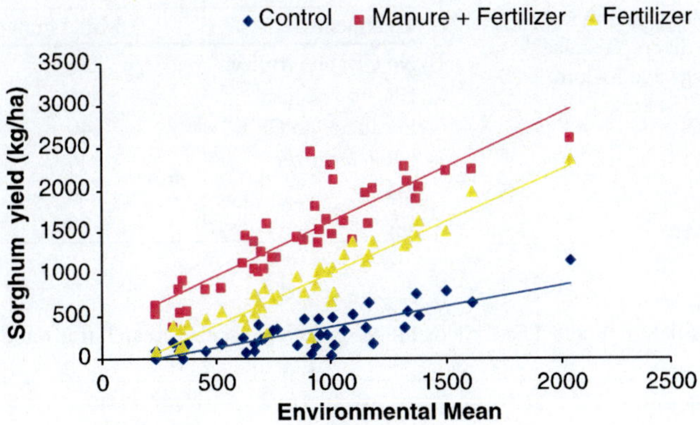

Fig. 1.2 Sorghum grain yield in relation to Environmental mean for a long-term trial (1960–2006) at Saria (Source Data: SARIA long-term (1960–2006) data)

limitations to decomposition. SOM depletion as compared to the fallow treatment seemed less pronounced in case of application of organic material with a relatively low C/N ratio such as manure. The adverse effect on soil organic matter and nitrogen status mostly affected the fraction of SOM >0.053 mm (Particulate Organic Matter, POM). The POM concentrations in the control, straw and urea-only plots were about one-half of the POM concentrations in the fallow plots. POM concentrations increased in the following order: urea < control < straw with or without urea < manure with or without urea < fallow. The fraction of SOM <0.053 mm (fine organic matter, FOM) was greater than POM in all plots except in fallow and manure + urea plots. Total nitrogen concentration followed the same trend as soil organic matter, but cultivation led to a decline in both POM-N and FOM-N. Crop yield was greatest in the manure plots and lowest in the straw, control and urea only plots. Results indicate that under Sudano-Sahelian conditions, SOM, POM and FOM fractions, fertilizer recovery fraction and crop performance were better maintained using organic resources with a low C/N ratio (manure) than through organic materiel with a relatively high C/N ratio (straw). Urea improved the effect of the organic material with high C/N ratio (straw) on crop yield and SOM concentration

Long-Term Crop Residue Management Trial (Sadore, Niger)

This randomized complete block design started in 1982 with main objectives being to assess soil fertility changes over time as affected by crop residue (CR) management. Later, a cereal-legume rotation was included in the system and each plot was split in two split-plots. The yield data that are available shows the advantage of combining organics (CR) and mineral fertilizers (Table 1.2). Application of both crop residues and fertilizer increase millet production from about 700 kg/ha to over 2,000 kg/ha compared to 96 kg/ha to about 400 kg/ha for the no input (control) treatment. Similar trend was observed for the millet totals dry matter.

Table 1.2 Millet grain yield in the long term crop residue management trial at Sadore, Niger, over 17 years

Treatments	Millet grain yield (kg/ha)
No CR, no fertilizer	285
CR, no fertilizer	591
Fertilizer, no CR	840
CR + fertilizer	1,344
SE	26
CV	28%

Table 1.3 Millet grain and TDM yields in the operational scale research trials at Sadore, Niger, 1998–2006

Treatments	Millet grain yield (kg/ha)	Millet TDM yield (kg/ha)
Traditional practices	129	943
Animal traction (AT) + no rotation +Intercropping + P	645	2,522
Animal traction (AT) + rotation +Intercropping + P	967	3,614
Hand cultivation (HC) + no rotation +Intercropping + P	546	2,414
Hand cultivation (HC) + rotation +Intercropping + P	871	3,502
Animal traction (AT) + no rotation+ Pure millet+P	639	2,796
Animal traction (AT) + rotation + Pure millet + P	998	4,343
Hand cultivation (HC) + no rotation + Pure millet + P	669	3,018
Hand cultivation (HC) + rotation + Pure millet + P	948	4,110
Traditional practices (local millet variety)	288	1,947

Results from this experiment showed that tillage using animal traction and hand cultivation resulted in significant increase in millet grain and biomass production (Table 1.3) compared to the traditional practices. This could be attributed to benefits of tillage such as improved water infiltration, soil aeration and root penetration. Systems involving crop rotation and intercropping had higher crop yields compared to those no rotation or intercropping. Rotation and intercropping systems involving cereals and legumes are known to result in improved nitrogen fixation as well as soil organic matter build up that can benefit current and subsequent crops.

Long-Term Effect of Manure, Crop Residue and Fertilizers (Sadore, Niger)

A factorial experiment was initiated in 1993 at a research station of ICRISAT Sahelian Center (ISC) at Sadore, Niger to assess the long-term effect of manure, crop residue and fertilizers on millet yields. Analysis of variance indicated that fertilizer, crop residue and manure application resulted in significant effect on both pearl millet grain and total dry matter yields. Of all the factors considered, fertilizer accounted for the highest total variation in the millet grain and dry matter. This was followed by manure.

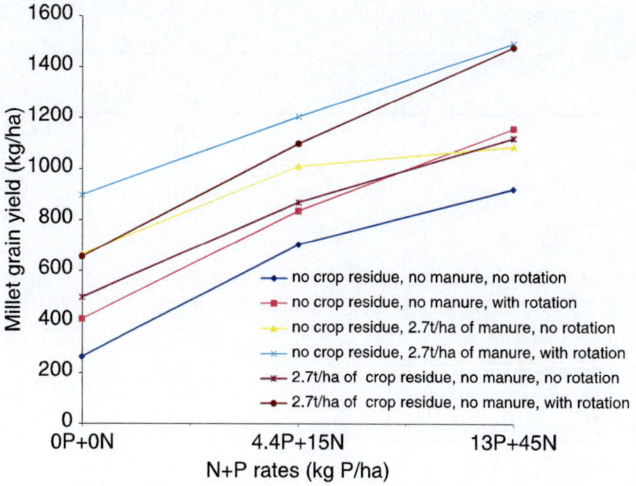

Fig. 1.3 Effect of fertilizer, CR, manure and rotation on pearl millet grain yield, Sadore, Niger, 1998–2006

Figure 1.3 illustrates the millet yield data over 9 years obtained. In this experiment, the farmer practice comprising of minimal crop residue retention, minimal manure application and no crop rotation yielded 262 kg/ha (Figure 1.3). However, application of 13 kg P/ha and 45 kg N/ha increased yields to 921 kg/ha. Combination of mineral fertilizer with manure and crop residues at the rate of 2.7t/ha each in a cowpea rotation system yielded over 1400kg of millet grain yield.

In 2001, the N and P fertilizer value of manure and crop residue was 27 and 13 respectively and the N and P equivalency of manure was 113% and 153% for crop residue (Table 1.4). These high values of fertilizers equivalencies of manure and crop residue over 100% suggest that the organic amendment have beneficial roles other than the addition of plant nutrient such as addition of micronutrients and better water holding capacity. In addition, the high fertilizer equivalencies could be due to improved synchrony between mineralization and plant nutrient uptake.

Effect of Degraded and Non Degraded Soils on Millet Production (Karabedji, Niger)

Soil degradation is a major problem in most of African farming systems. A trial was established in Karabedji, Niger in 1999 to assess the effect of degraded and non degraded soils on millet production (Karabedji, Niger). Mineral fertilizers were applied on two major type of soils: Farms close to the village where household waste, human excreta and farm yard manure are commonly used were considered as non-degraded fields while farms far from the village and which received no organic material added were considered as the degraded fields. Results from this experiment

Table 1.4 Fertilizers equivalency of manure and crop residue at Sadore, Niger, 2001 cropping season

Parameters	Grain (kg/ha)	Total dry matter (kg/ha)
Absolute control	236	1,523
% N in manure	0.71	0.71
% P in manure	0.18	0.18
% N in crop residue	0.71	0.71
% P in crop residue	0.03	0.03
Yield at 2.7 t/ha of manure in continuous cropping	800	3,461
Yield at 2.7 t/ha of crop residue in continuous cropping	634	2,527
Equivalent N and P of the manure	27	58
Equivalent N and P of the crop residue	13	58
Fertilizer N and P equivalency of manure (%)	113	243
Fertilizer N and P equivalency of crop residue (%)	153	290

showed higher fertilizer P response in home gardens as compared to the away farms. These results support the observation that fertilizer response are higher in fertile than in degraded soils.

Long Term Trial at Kouaré, Burkina Faso

A study was conducted in Kouaré, Burkina Faso to assess the contributions of cowpea and fallow to soil fertility improvement (Bado 2002). Five rotation treatments (cowpea-sorghum, sorghum-cowpea, fallow-sorghum, sorghum-fallow and sorghum-sorghum) were used as first factor and four fertilizer treatments (chemical Nitrogen (N), Phosphorous (P) and Potassium (K) fertilizer; NPK + Dolomite; P + Manure and Control without any fertilizer) were used as second factor

A overall analysis of data of the 5 years indicated that all fertilizer treatments improved sorghum grain and stover yields (Table 1.5). Chemical NPK fertilizer improved sorghum grain yields but higher yields were obtained when chemical NPK fertilizer was associated with dolomite. The application of chemical P fertilizer (in lack of N and K) associated with manure was as effective on sorghum grain yield as chemical NPK fertilizer alone.

Sorghum produced less than 500 kg/ha of grain in mono cropping but produced highest grain yields when fallow or cowpea was used in the cropping system (fallow-sorghum or cowpea-sorghum rotation). The succeeding sorghum grain yields increased from 75% to 100% when it was rotated with fallow and cowpea respectively. The effects of rotations increased over time and cowpea was most effective on sorghum yield increase during the last 3 years (Fig. 1.4). Total N uptake by succeeding sorghum increased from 26 kg N/ha in mono cropping of sorghum to 31 and 48 kg N/ha when sorghum was rotated with fallow or cowpea respectively (Bado 2002; Bado et al. 2006) (Table 1.6).

Table 1.5 Sorghum and cowpea yields (kg/ha) as affected by fertilization and cropping systems during 4 years (1996–1999) at Kouaré, Burkina Faso

	Sorghum		Cowpea	
Fertilization	Grain	Stover	Grain	Haulms
NPK+ Dolomite	914[a]	4,489[a]	611[ab]	1,228[a]
P+ manure	746[bc]	3,769[c]	649[a]	1,284[a]
NPK	798[b]	4,205[ab]	510[c]	1,028[a]
Control	346[d]	1,611[d]	265[d]	380[b]
Cropping systems				
Cowpea-Sorghum	938[a]	4,281[a]	na	na
Fallow-Sorghum	819[b]	3,493[b]	na	na
Sorghum-Sorghum	467[c]	2,912[c]	na	na

Values followed by the same letter in the same column are not significantly different at $p<0.05$, according to Fisher's test
na Not applicable

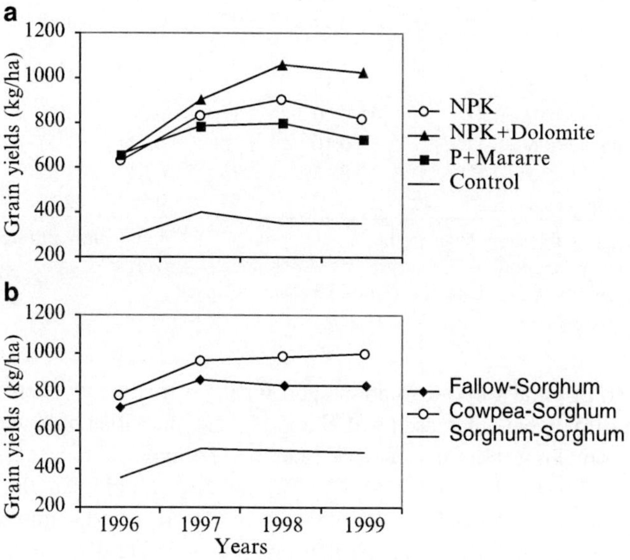

Fig. 1.4 Effects of (**a**) fertilizer applications and (**b**) crop rotations on succeeding sorghum grain yields during 4 years (1996–1999) at Kouaré, Burkina Faso

After 5 years of cultivation (1994–1999), soil organic carbon was significantly affected by crop rotation ($p<0.05$), but fertilization did not affect soil organic C (Table 1.7). The highest concentration of organic C was observed in soils of fallow-sorghum rotation and lowest quantities were observed in mono cropping of sorghum and cowpea-sorghum rotation. As observed with organic C, the highest quantity of N was observed in soils of fallow-sorghum rotation and lowest quantities were observed in mono-cropping of sorghum and cowpea-sorghum rotation.

Table 1.6 Effects of cowpea and fallow on total N uptake by succeeding sorghum, fertilizer N use efficiencies (FNUE) and N derived from soil (Ndfs) and fertilizer (Ndfs) in 1999 after 5 years of cultivation

Cropping systems	Total N uptake, kg N/ha	FNUE, %	Ndff, kg/ha	Ndfs kg/ha
Fallow-Sorghum	31[b]	26[a]	10[a]	21[b]
Cowpea-Sorghum	48[a]	22[ab]	8[ab]	40[a]
Sorghum-Sorghum	26[c]	17[c]	6[bc]	20[bc]

Values followed by the same letter in the same column are not significantly different at $p<0.05$, according to Fisher's test

Table 1.7 Some soil (0–20 cm layer) properties as affected by fertilizers and crop rotations after 5 years of cultivation (1995–1999)

	pH KCl	C. org (%)	Total N Kg/N	K+ Cmol/kg	Ca++ Cmol/kg	Mg++ Cmol/kg	ECEC Cmol/kg	Bases (%)	Al (%)
Cropping systems									
Fallow-Sorghum	5.3	0.39[a]	360[a]	0.11	1.43[a]	0.44	2.10	95	3
Cowpea-Sorghum	5.2	0.29[b]	294[b]	0.09	1.09[b]	0.53	1.84	94	3
Sorghum-Sorghum	5.3	0.29[b]	335[b]	0.14	1.07[b]	0.35	1.68	94	3
Fertilization									
P+Manure	5.2[b]	0.32	318[a]	0.10	1.15	0.33	1.72	93[b]	4[b]
NPK	5.0[c]	0.33	317[a]	0.10	1.12	0.27	1.70	88[c]	6[a]
NPK+Dolomite	5.6[a]	0.32	270[b]	0.10	1.38	0.62	2.17	99[a]	0[d]
Control	5.3[b]	0.32	237[c]	0.15	1.13	0.54	1.91	97[a]	2[c]
Original soil	5.5	0.50	427	0.13	1.59	0.51	2.30	99	0

Values followed by the same letter in the same column are not significantly different at $p<0.05$, according to Fisher's test
EA Exchange acidity, *ECEC* Effective Cation Exchange Capacity

Compared to original soil, all rotations and fertilizer treatments decreased soil N. All fertilizer treatments increased soil N and highest quantities of N were observed in manure containing treatment (PK + manure). Dolomite increased soil pH and maintained soil bases (Ca^{++} and Mg^{++}), ECEC and Al^{3+} saturation at the same levels as the original soil. Soil pH and bases saturation were significantly greater with combined application of P and manure than chemical NPK fertilizer alone. Chemical NPK fertilizer increased Al^{3+}. But manure application decreased Al^{3+} saturation.

Long-Term Fertility Experiments (LTFE) for Irrigated Rice in the West African Sahel: Agronomic Results

The LTFE were started at WARDA's two research farms in Ndiaye and Fanaye, Senegal, in the 1991 (Haefele et al. 2002; Bado et al. 2010). The intention was to monitor changes in soil fertility and rice yield response to N, P and K fertilizer input in the intensive monocropping system of irrigated rice. The trial consists of six treatments.

Table 1.8 Average grain yields in the long-term fertility experiment at Ndiaye and Fanaye during 20 continuous seasons between 1991 and 2001 for the wet season (WS) and the hot dry season (HDS)

Treatment	N-P-K doses (kg/ha)	Ndiaye WS (t/ha)	Ndiaye HDS (t/ha)	Fanaye WS (t/ha)	Fanaye HDS (t/ha)
T1	0-0-0	3.2	3.5	3.7	2.0
T2	120-26-50	6.7	7.0	6.9	6.5
T3	120-52-100	6.5	6.8	7.1	6.5
T4	120-0-0	5.3	5.3	5.8	4.4
T5	180-26-50	6.6	7.5	7.6	7.6
T6	60-26-50	5.6	5.8	5.8	4.9
Average		**5.7**	**6.0**	**6.2**	**5.3**

Average grain yield of all treatments was highest in the Fanaye WS (6.2 t/ha), intermediate in the Ndiaye WS and HDS (6.0 and 5.7 t/ha, respectively), and lowest in the Fanaye HDS (5.3 t/ha), (Table 1.8). T1 had always the lowest yield (2.0–3.7 Mg/ha) and was always significantly different from all other treatments. T5 resulted in the highest yields (6.6–7.6 Mg/ha) except for the WS in Ndiaye.

Long Term Trials in East Africa

Long Term Experiment, Kabete, Kenya

The Long term experiment at the National Agriculture Research laboratories (NARL) Kabete, Kenya was established in 1976 to study small-scale farmers' methods of improving and sustaining soil fertility (Kibunja and Gikonyo 2000). The primary objective of the trial was to find appropriate methods for maintaining and improving the productivity of the soils through repeated use of inorganic fertilizers (in particular nitrogen and phosphorus), farm yard manure and crop residues under continuous cropping following the husbandly practices of small-scale farmers. This sections presents only a summary of the key lessons and more detailed results from this experiment are discussed in Chapter 4.

At the onset of the experiment, yields averaged slightly over 3.5 t/ha. However, great variations have been recorded due to treatment effects, rainfall amounts and distribution as well as timeliness in planting. In general, application of chemical fertilizers gave higher yields than FYM alone for the first 6 years. However, after the tenth year, maize yields for NP treatments were significantly lower than for FYM alone. During the next 10 year period of continuous cropping, the differences between FYM and FYM+NP combination became increasingly minimal and the yields tended to decline more rapidly where only chemical fertilizers were applied. To date the results indicate that 5–10 t/ha of FYM applied singly or in combination with 60 kg N and 60 kg P_2O_5 chemical fertilizers is the most promising nutrient

Table 1.9 Long term mean maize grain yield (t/ha) for selected treatments at Embu, Kenya (1993–2005)

Treatment	Mean Grain Yield (t/ha)
Calliandra alone	2.7
Leucaena alone	2.7
Calliandra+30kg N/ha	3.1
Leucaena+30kg N/ha	3.1
Recommended fertilizer rate (60 kg N/ha)	2.3
Control	1.6

management strategy for long-term improvement and maintenance of soil fertility. Other studies on this experiment have shown that there was a general reduction in soil organic matter (SOM) across all the treatments. However, SOM fractions were more favorably influenced by the addition of FYM than stover retention (Kapkiyai et al. 1998, 1999).

Hedgerow Intercropping Trial in Embu

Recognizing the potential of leguminous trees species to improve soil fertility, the Hedgerow Intercropping (HI Embu) trial was initiated in 1992 at the Kenya Agricultural Research Institute (KARI) Regional Research Centre (Embu) in the central highlands of Kenya (Mugendi et al. 1999). The experiment sought to investigate feasibility of using leguminous trees for soil fertility improvement. The leguminous shrubs used were *Leucaena leucocephala* (Lam.) de Wit and *Calliandra calothyrsus* Meissn applied as biomass transfer (from outside) or as insitu (from the hedgerow prunnings). The trial consists of ten treatments set out in a randomized complete block design (RCBD) with four replicates.

The average maize grain yields obtained across the 11 years (22 seasons) indicated that treatments receiving both organic and inorganic nutrient sources recorded the highest yields that were compareable to the continous fertilizer application treatment (Table 1.9). Leucaena alley cropped treatment with prunings incorporated (Treatment 2) gave an average yield of 2.8 t/ha whereas the equivalent calliandra treatment (Treatment 1) recording an average yield of 2.0 t/ha. The above yields are compared to the recommended fertilizer and the control treatment yields of 2.3 and 1.6 t/ha respectively.

Treatments with prunings incorporated with fertilizer gave better maize grain yields compared to treatments with only prunings applied. This could be due to the provision of additional physical, chemical and biological benefits (besides N) by the organic materials or due to the prevention of other nutrients deficiencies and/or enhanced nutrient fluxes (Sanchez and Jama 2002).

Kirege Experiment, Eastern Kenya

Kirege trial, in Chuka location, Meru South District Eastern Kenya was established in March 2000 to evaluate the potential of various soil fertility management technologies

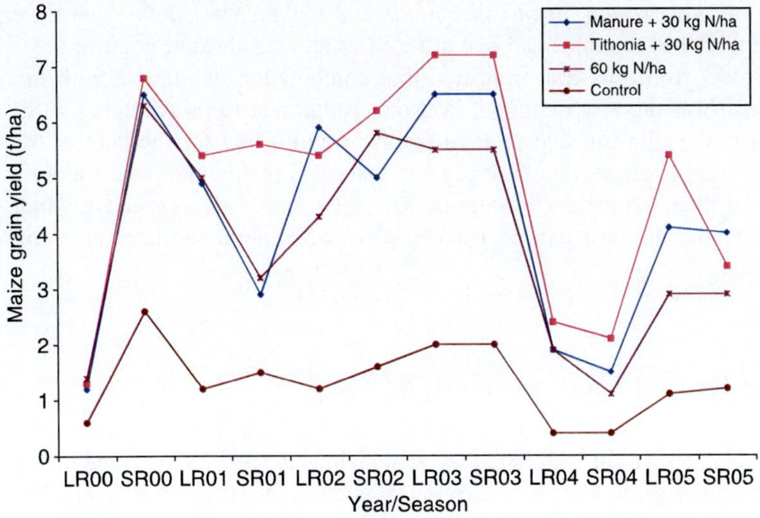

Fig. 1.5 Maize yields under different technologies in Chuka, Meru South District during the period 2000 to 2005

to improve soil fertility. The experiment was laid out as a randomized complete block design (RCBD) with the plot sizes measuring 6 m×4.5 m replicated thrice. Nine external soil fertility amendment inputs were applied every season to give an equivalent of 60 kg N/ha, the recommended rate of N application to meet maize nutrient requirement for an optimum crop production in the area (FURP 1987). These treatments were compared to an absolute control where no external inputs were applied.

The average maize grain yields in the different treatments across the 12 seasons are presented in Fig. 1.5. The maize grain yields monitored across 12 seasons (6 years) were significantly different ($p<0.05$) among treatments. On average, treatments involving a combination of organics and inorganic fertilizer resulted in higher yields compared to the sole fertilizer or the absolute control treatment (Fig. 1.5).

This could be due to the provision of additional physical, chemical and biological benefits (besides N) by the organic materials or due to the prevention of other nutrients deficiencies and/or enhanced nutrient fluxes (Sanchez and Jama 2002). The control gave consistently low yields across the 12 seasons and it reflects to a large extent the yields attained on most farmers' fields where very low inputs or none are applied.

Nitrogen Fertilizer Equivalence Experiment Kabete, Kenya

The Nitrogen Fertilizer Equivalence Experiment (N1 Kabete) was established in 1999 with the objective of assessing the fertilizer equivalencies of the organic materials used for soil fertility management (Kimetu et al. 2004). Three organic resources

(*Tithonia diversifolia, Senna spectabilis* and *Calliandra calothyrsus*) were compared with the recommended rate of fertilizer and the absolute control.

Results from this trial indicated that combination of tithonia with inorganic fertilizer was the best treatment over the cropping seasons averaging 7 t/ha compared to 6.5 t/ha for sole mineral fertilizer and 4 t/ha for the control (data not shown). On average, the three organic materials (calliandra, senna and tithonia) gave fertilizer equivalency values of 50%, 87% and 118%, respectively indicating that tithonia biomass can be used in place of mineral fertilizer as a source of nitrogen.

Long Term Trials in Southern Africa

Some long term trials were established in Malawi in 1995/1996 to predict long-term changes in organic C, N and texture under different cropping systems. The trials were initially established at five sites: Chitala, Chitedze, Lisasadzi, Makoka and Zombwe.

The design was a split plot with cropping systems allocated in main plots and residue management in subplots. Factors under study were three cropping systems with or without fertilizer and two crop residue management options (Table 1.10).

Initial site characteristics are indicated in the Table 1.11. Soil pH ranged from 5.3 (Zombwe and Makoka) to 5.6 (Chiatala). Soil pH is in the good range to support crop production. All sites had low initial soil nitrogen that made all sites suitable for this study. Initial soil nitrogen ranged from 0.002% to 0.16%. Sites had different levels of soil organic carbon ranging from 0.30% to 2.07%.

Average annual total rainfall varied across years and sites (Fig. 1.6). More critical was the high monthly variations which often occurred during planting and flowering and grain filling stages (January and February). Dry years were observed in 1995, 1999, 2000, 2001, 2004 and 2005.

Maize grain yield recorded from the four sites differed significantly due to treatment effects, site differences in soil types and climatic conditions (Figs. 1.7 and 1.8). Despite differences in the sites, the response trend in maize yields to different cropping systems remained the same. The highest grain yield across sites were obtained when maize were grown in rotation with pigeon pea followed by maize/pigeon pea intercropping. The maize/pigeon pea rotation treatment gave similar yields in all the sites when compared at the same fertility level. This shows that pigeon pea had the largest nitrogen residual effect to benefit the subsequent maize crop.

Highest maize yields were obtained in the initial year across sites of the trial as indicated in (Figs. 1.9 and 1.10). Average maize yields decreased with time. Deflection in maize yields in some years and sites was attributed to drought conditions and change of the trial variety index from MH18 to DK3031. The initial maize variety index MH 18 was changed because the former variety became scarce on the market due to its high susceptibility to foliar diseases such as Grey Leaf Spot. Normally DK 8031 has a higher yield potential than MH18.

Table 1.10 Main and subplot treatment description at Chitedze, Chitala, Makoka and Zombwe

Plot	Fertility	Cropping system	Residue management	
		Main plot	Subplot 1	Subplot 2
1	No fertilizer added	Sole maize	Minus stover	Plus stover
2		Maize/pigeon pea rotation	Minus pigeon pea stems + pigeon pea leaves	Plus pigeon pea stems + pigeon pea leaves
3		Maize/pigeon pea intercropping	Minus p/pea stems +p/pea leaves	plus maize residues + p/pea leaves + p/pea stems
4	Fertilizer added (area specific recommendation)	Sole maize	Minus stover	Plus stover
5		Maize/pigeon pea rotation	Minus pigeon pea stems + pigeon pea leaves	Plus pigeon pea stems + pigeon pea leaves
6		Maize/pigeon pea intercropping	Minus p/pea stems +p/pea leaves	plus maize residues + p/pea leaves + p/pea stems

Table 1.11 Initial soil physical and chemical characteristics of experimental sites

Site	pH (H_2O)	OC	N	Clay	Silt	Fine sand	Coarse sand	Soil texture
		Percentage (%)						
Chitala	5.6	1.07	0.07	26	9	25	40	SCL
Chitedze	5.4	2.39	0.16	28	21	21	30	SCL
Lisasadzi	5.8	0.59	0.04	4	4	36	56	SL
Makoka	5.3	0.30	0.02	16	6	28	50	LS
Zombwe	5.3	0.68	0.05	22	4	30	44	SCL

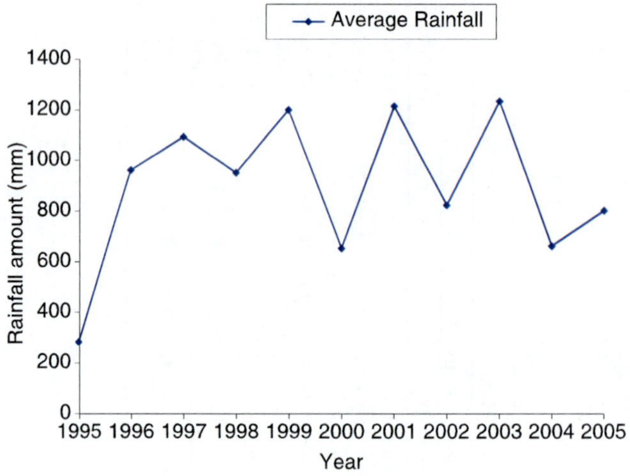

Fig. 1.6 Rainfall means (mm) across years for 10 years (1995–2005)

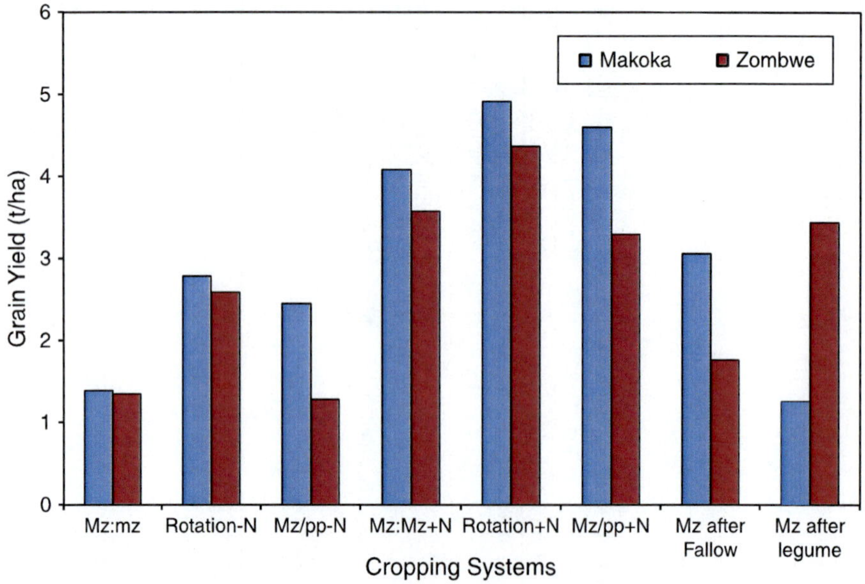

Fig. 1.7 Maize yield response to different cropping systems over a period of 5 years (2000–2005) at Makoka and Zombwe

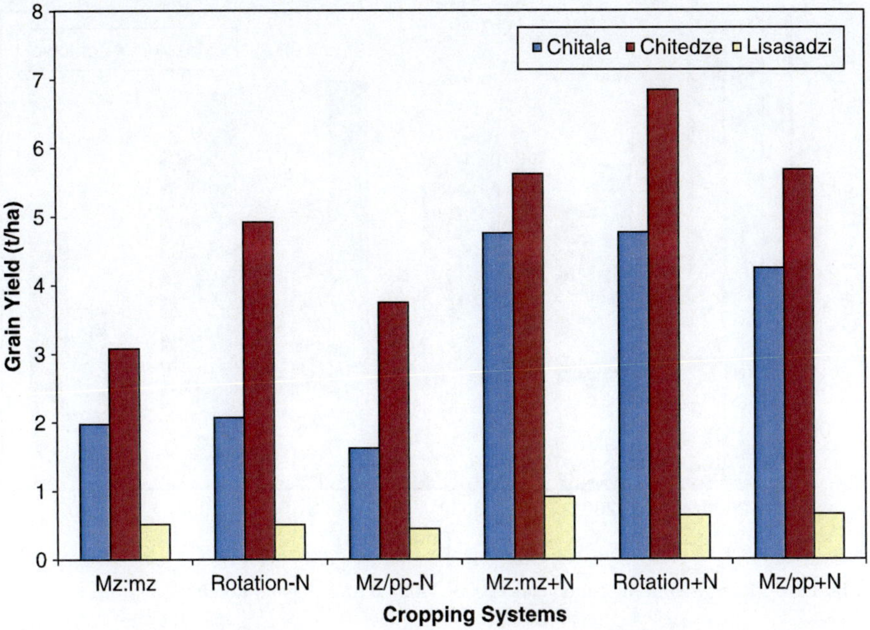

Fig. 1.8 Maize yield response to different cropping systems over a period of 6 years (2000–2005) at Chitala, Chitedze and Lisasadzi

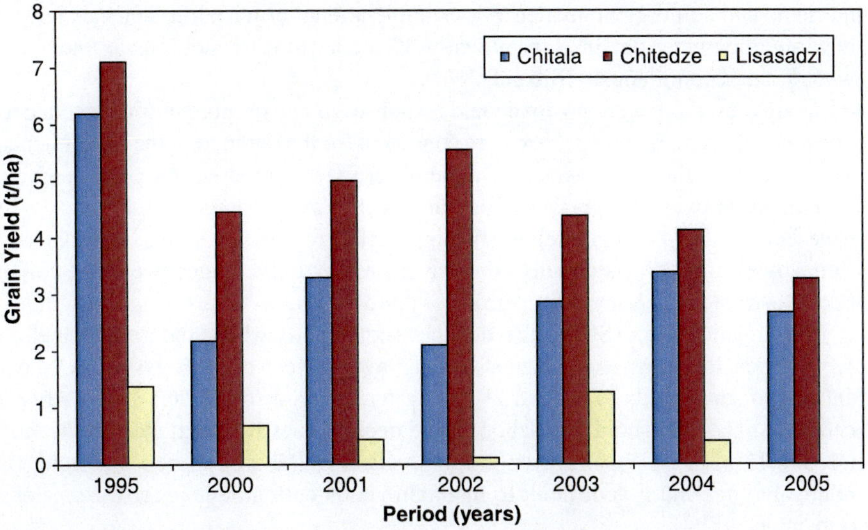

Fig. 1.9 Maize yield trend over a period of 7 years (1995–2005) at Chitala, Chitedze and Lisasadzi

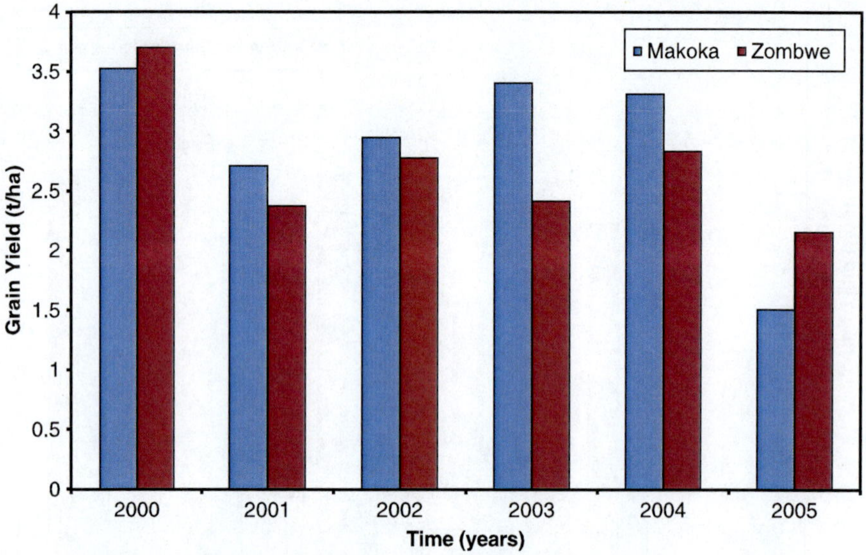

Fig. 1.10 Maize yield trend for 6 years (2000–2005) at Makoka and Zombwe

Key Lessons from Long-Term Experiments in Africa

A review of these experiments highlights a number of key findings:

All long term trials showed yield decline, often with a relatively rapid fall to a low level equilibrium. Such yield dynamics are highly correlated with available nutrients and soil organic matter. For example a long term trial in Senegal showed how organic matter declined from nearly 3% under forest conditions to under 1% when farmed continuously (Siband 1974).

At all sites, there were positive yield responses to one or more nutrients added as mineral fertilizers. The responses were consistent for the duration of the experiments. This highlights the effectiveness of mineral fertilizers in increasing yield in arable farming systems in Africa. This potential is recognized by largescale farmers, who have been able to sustain relatively high yields of maize (Kenya, Zambia and Zimbabwe), tobacco (Nicotiana tabacum L.; Malawi and Zimbabwe), and coffee (Coffea Arabica L.; Kenya) for periods of up to 30 years.

Soil organic matter (SOM) also declines significantly when land is cultivated. In West Africa, long term experiments show a range between over 5% loss of SOM per annum on sandy soils to around 2% on better textured soils (Pieri 1995) After a certain point, a threshold is reached where decline does not continue significantly (around 1%, Lal 1995). However, at low levels of SOM, crop response to inputs is relatively poor and it is difficult to maintain yields with inorganic fertilizers alone (Greenland 1994).

Prolonged treatments using only inputs of organic matter (animal manure, green manure, crop residues, etc.) also showed yield declines, although the positive impact were sustained longer than for inorganic fertilizers alone in many cases. However, the amount of organic matter required to sustain yields was considerable, with very large application rates needing to be applied in many experiments.

Rotational treatments, including sequences with legume crops and fallow periods had lower yield declines than monocultures. Such treatment also had lower rates of SOM loss (Pieri 1995). Most showed that reducing tillage frequency, including no-till options, reduced the rates of organic matter loss (Pieri 1995), although the impact on overall yield was often mixed.

The best results (in term of long-term sustained yield response) invariably were those treatments that combined inorganic and organic inputs. For example, over 18 years in Kabete, in the favorable highland conditions of Kenya, maize consistently yielded highest under the combined treatment. This was attributed to the achievement of the appropriate nutrient input (most easily through the choice of appropriate inorganic fertilizers mixes) with good soil structure (through effective organic matter management).

Limitations of the Long Term Trials

All of the long term experiments were located on research stations and were managed by researchers. Although early experiments tended to look at a variety of treatments, including rotations, green manuring, and animal manuring, later experiments shifted to look more specifically at inorganic fertilizers, and failed to capture the range of farmer's practices.

Only a few of the long term experiments in Africa spanned a period of longer than 20 years, making it difficult to assess longer term dynamics. Termination of these trials has been mainly due to lack of funds to maintain the experiments.

Relatively few of these experiments provide details of soil changes, so they can not be used to test for patterns of soil fertility decline or nutrient removal over time. Other factors that have limited interpretation of these time-series data include variations in the inter-annual, seasonal and cyclical changes in rainfall or pest and disease incidences.

Results from long-term trials have not been used satisfactorily to influence policy or management recommendations.

The Way Forward

Results obtained from short-term experiments cannot provide the appropriate scientific information and could produce misleading technological recommendations. Long term trials hold key to understanding the processes and functioning of many of the cropping systems in tropical Africa. These trials could be used especially by researchers and higher education students interested in basic research about the various systems. Data from these trials can be used for simulation and prediction of potential changes in production or soil properties as thus improving on decision making process to adapt to the changing conditions. Despite these importance, only a handful of such trials with lifespans of over 20 years exist across the continent. Those in existence have faced problems of inconsistency in management and data collection. In addition, most of these trials lack comprehensive one-stop databases hence it is difficult to monitor their performance.

Long-term experiments are costly to maintain over time, as they require constant and valid management as well as accuracy and adequacy in data collection and management (Pieri 1995). This financial burden may be considered excessive particularly by national, regional or even international research organizations working in developing countries which prefer to address the most pressing agronomic issues with short-term trials. Long term funding is required to maintain long term trials in Africa. This can be achieved through national support and from collaborative research leading to MSc and PhD with various local and foreign research institutions and universities.

References

Bado BV (2002) Rôle des légumineuses sur la fertilité des sols ferrugineux tropicaux des zones guinéennes et soudaniennes du Burkina Faso. Thèse de Doctorat (PhD) Département des sols et Environnement, Université Laval, Quebec, 145 p

Bado BV, Bationo A, Lompo F, Segda Z, Cescas MP, Sedogo MP (2006) Long-term effects of cropping systems and fertilization on crop production, soil characteristics and nitrogen cycling in the Guinean and Sudanian savannah zones of Burkina Faso (West Africa). In: Zapata F, Nguyen L-M (eds) Management practices for improving sustainable crop production in tropical acid soils. Results of a coordinated research project organized by the Joint FAO/IAEA Programme of Nuclear Techniques in Food and Agriculture, IAEA Proceedings Series. International Atomic Energy Agency, Vienna, pp 47–64

Bado BV, Aw A, Ndiaye M (2010) Long-term effect of continuous cropping of irrigated rice on soil and yield trends in the Sahel of West Africa. In: Bationo A, Waswa, BS, Okeyo J, Maina F, Kihara J (eds) SI: Innovations as Key to the Green Revolution in Africa: Exploring the Scientific Facts. Nutr Cycl Agroecosyst 88:133–141

Bekunda MA, Bationo A, Ssali H (1997) Soil fertility management in Africa: a review of selected research trials. In: Buresh RJ, Sanchez PA, Calhoun F (eds) Replenishing soil fertility in Africa. SSSA Special Publication No. 51. SSSA, Madison

CGIAR (2001) Special panel on impact assessment, Contributions Made by the CGIAR and Its Partners to Agricultural Development in Sub-Saharan Africa. Consultative Group on International Agricultural Research, Washington, DC

DFID (2000) Sustainable livelihoods guidance sheets. The Department for International Development, London

FURP (1987) Description of the first priority trial sites in various districts. KARI, FURP, Nairobi, Kenya, pp 1–31

Greenland DJ (1994) Long-term cropping experiments in developing countries: the need, the history and the future. In: Leigh RA, Johnson AE (eds) Long-term experiments in agricultural and ecological sciences. CAB International, Cambridge

Haefele SM, Wopereis MCS, Wiechmann H (2002) Long-term fertility experiments for irrigated rice in the West African Sahel: agronomic results. Field Crop Res 78(2002):119–131

Kapkiyai JJ, Karanja NK, Woomer PL, Qureshi JN (1998) Soil organic carbon fractions in a long-term experiment and the potential for their use as a diagnostic assay in highland farming systems of Central Kenya. Afr Crop Sci J 6(1):19–28

Kapkiyai JJ, Karanja NK, Qureshi JN, Smithson PC, Woomer PL (1999) Soil organic matter and nutrient dynamics in a Kenyan nitisol under long-term fertilizer and organic input management. Soil Biol Biochem 31(13):1773–1782

Kibunja C, Gikonyo E (2000) Maintenance of soil fertility under continuous cropping in a maize-bean rotation (Long-term trial, NARL-Kenya). TSBF Report 1997–1998. UNESCO, Nairobi

Kimetu JM, Mugendi DN, Palm CA, Mutuo PK, Gachengo CN, Bationo A, Nandwa S, Kungu JB (2004) Nitrogen fertilizer equivalencies of organics of differing quality and optimum combination with inorganic nitrogen source in Central Kenya. Nutr Cycl Agroecosyst 68(127–135):2004

Lal R (1995) Technological options for sustainable management of Alfisols and Ultisols in Nigeria. In: Lal R, Stewart BA (eds) Soil management: experimental basis for sustainability and environmental quality. Lewis Publishers, London

Lynam JK et al (1998) Nutrient balances as indicators of productivity and sustainability in sub-Saharan African agriculture introduction. Agric Ecosyst Environ 71:1; Drechsel P, Gyiele LA (1999) The economic assessment of soil nutrient depletion. International Board for Soil Research and Management, Bangkok

Mando A (2006) Effect of long-term addition of organic input of contrasting quality on soil carbon and crop performance under Sudano-Sahelian conditions. 18th World Congress of Soil Science, Philadelphia, Pennsylvania, 9–15 July 2006

Mando A, Bonzi M, Wopereis MCS, Lompo F, Stroosnijder L (2005) Long-term effects of mineral and organic fertilization on soil organic matter fractions and sorghum yield under Sudano-Sahelian conditions. Soil Use Manag 21:396–401

Mugendi DN, Nair PKR, Mugwe JN, O'Neill MK, Woomer PL (1999) Alley cropping of maize with calliandra and leucaena in the subhumid highlands of Kenya. Part 1. Soil-fertility changes and maize yield. Agrofor Sys 46(1):39–50

Pieri C (1995) Long term management experiments in semiarid Francophone Africa. In: Lal R, Stewart BA (eds) Soil management: experimental basis for sustainability and environmental quality. CRC Press, Boca Raton, pp 225–266

Pinstrup-Andersen P, Pandhya-Lorch R, Rosengrant MW (2000) World food prospects: critical issues for the early 21st century (1999) and food security: problems, prospects and policies (2000). International Food Policy Research Institute, Washington, DC; available at www.ifpri.org

Sanchez PA (1994) Tropical soil fertility research: towards the second paradigm. In: Inaugural and state of the art conferences. Transactions 15th World Congress of Soil Science. Acapulco, Mexico, pp 65–88

Sanchez PA (2002) Soil fertility and hunger in Africa. Science 295(5562):2019–2020

Sanchez PA, Jama BA (2002) Soil fertility Replenishment Takes off in East and Southern Africa. In: Vanlauwe B, Diels J, Sanginga N, Merckx R (eds) Integrated plant nutrient management in Sub-Saharan Africa: from concept to practice. CAB International, Wallingford, Oxon

Sanchez PA, Shepherd KD, Soule MJ, Place FM, Buresh RJ, Anne-Marie N, Izac A, Mokwunye U, Kwesiga FR, Ndiritu CG, Woomer PL (1997) Soil fertility replenishment in Africa: an investment in natural resource capital. In: Buresh RJ, Sanchez PA, Calhoun F (eds) Replenishing soil fertility in Africa. SSSA Special Publication No. 51 Soil Science Society of America. Madison, Wisconsin, pp 1–46

Siband P (1974) Evolution des caractéres et de la fertilité d'un sol rouge de Casamance. L'Agronomie Tropicale 29:1228–1248

Smaling EMA (1993) An agro-ecological framework of integrated nutrient management with special reference to Kenya. Wageningen Agricultural University, Wageningen

Southwood TRE (1994) The importance of long-term experimentation. In: Leigh RA, Johnson AE (eds) Long-term experiments in agricultural and ecological sciences. CAB International, Cambridge

Swift MJ, Seward PD, Frost PGH, Qureishi JN, Muchena FN (1994) Long-term experiments in Africa: developing a database for sustainable land use under global change. In: Leigh RA, Johnson AE (eds) Long-term experiments in agricultural and ecological sciences. CAB International, Cambridge

Theng BKG (1991) Soil science in the tropics – the next 75 years. Soil Sci 151:76–90

Chapter 2
Long Term Effects of Crop Rotations with Fallow or Groundnut on Soil Fertility and Succeeding Sorghum Yields in the Guinea Savannah of West Africa

B.V. Bado, A. Bationo, F. Lompo, K. Traore, M.P. Sedogo, and M.P. Cescas

Abstract The effects of crop rotations with fallow, groundnut (*Arachis hypogea*) and cotton (*Gossypium sp*) on soil and subsequent sorghum (*Sorghum bicolour*) yields were studied with a 14-years (1993–2007) old experiment at Farakô-Ba (4° 20′ West, 11° 6′ North and 405 m altitude) in the Guinean z one of Burkina Faso. A factorial design in a split plot arrangement with four rotation and eight fertilizer treatments were used.

Crop rotations with groundnut or annual fallow increased succeeding sorghum yields. Mean annual yields of succeeding sorghum increased from 547 kg ha^{-1} in continuous sorghum to 912 and 1,021 kg ha^{-1} in Fallow-Sorghum and Cotton-Groundnut-Sorghum rotations, respectively. Soils of Fallow-Sorghum and Cotton-Groundnut-Sorghum rotations released more mineral Nitrogen (N) at sowing

B.V. Bado (✉)
INERA, CRREA de l'Ouest, Farakô-Ba, PB 910, Bobo-Dioulasso, Burkina Faso

Africa Rice Center (AfricaRice), Sahel Regional Station BP 96, Saint-Louis, Senegal
e-mail: V.Bado@cgiar.org

A. Bationo
AGRA/West Africa, CSIR Office Complex #6 Agostino Neto Road, PMB KIA 114
Airport – Accra, Ghana

F. Lompo • M.P. Sedogo
INERA, 04 BP 8645 Ouagadougou, Burkina Faso

K. Traore
INERA, CRREA de l'Ouest, Farakô-Ba, PB 910, Bobo-Dioulasso, Burkina Faso

M.P. Cescas
Département des Sols et Génie Agroalimentaire (FSSA), Université Laval Ste Foy
Québec (QC), 2219 Comtois, QC, Canada G1K 7P4

A. Bationo et al. (eds.), *Lessons learned from Long-term Soil Fertility Management Experiments in Africa*, DOI 10.1007/978-94-007-2938-4_2,

and increased fertilizer N use efficiency from 13 to 32 units, respectively, compared to continuous sorghum. Soil organic carbon increased from 0.36% in continuous sorghum to 0.39% and 0.54% in Cotton-Groundnut-Sorghum and Fallow-Sorghum rotations respectively. Except for fallow-sorghum, all rotations decreased soil organic carbon, soil pH with an increase of Aluminium ion (Al^{3+}) saturation. Highest quantities of extractable Phosphorus (P) were observed in mono-cropping of sorghum. Manure increased soil organic carbon (SOC), total N and available P. Manure or dolomite decreased Al^{3+} saturation.

Keywords Crop rotations • Fertilization • Fallow • Groundnut • Nitrogen • Soil

Introduction

Crop yields in most agricultural systems of West Africa are very low. Despite their initial good agricultural productivity, crop yields decrease over time because of low nutrient inputs addition and consequent soil fertility decline (Pichot et al. 1981; Pieri 1989; Bado et al. 1997). Livestock is not well integrated with agricultural activities and crop residues are usually exported from the farm for household needs and animal feeding. In traditional systems, soil fertility maintenance was based on a relatively long fallow period (10–15 years of fallow) followed by a short cropping period (3–5 years) (Steiner 1991). However in recent years, increased population pressure has resulted in significant changes of the traditional bush-fallow systems. Long-term fallow systems are no longer practical and lands are continuously cultivated for long time leading to soil fertility decline (Bationo and Mokwunye 1991b). Despite the recognized need to apply chemical fertilizers to maintain soil fertility and obtain high yields, their use in West Africa is limited by a number of factors such as uncertainty of climatic conditions, in particular high drought risk, land tenure, lack of financial resources, inefficient distribution systems, poor agricultural production and marketing-supporting policies and other socio-economic factors. Alternatively, cost-effective and socio-economic viable means of improving and maintaining soil fertility and productivity must therefore be developed and pilot-tested before providing recommendations to the farmers. These include increasing the efficiency of the little applied fertilizer and enhancing the nutrient inputs addition and recycling from other sources at the level of cropping systems.

In the Guinean savannah zone of Burkina Faso, sorghum, maize and cotton are the main basic crops cultivated by farmers. N_2-fixing legume crops such as groundnut (*Arachis hypogea L.*) and cowpea (*Vigna unguiculata*) are also important components of the cropping systems for the resource-poor farmers because they serve as food as well as cash crops. They can eventually provide necessary financial resources to buy agricultural inputs, for instance fertilizers. Rotation of cereals with legumes has been studied in recent years and beneficial rotational effects on subsequent cereal yields have been reported for the Sahelian zone of West Africa (Bationo and Ntare 2000; Bagayoko et al. 2000). The main effect of legumes on succeeding crops is commonly attributed to an increase in soil N fertility as a result of biological N_2 fixation (Dommergues and Ganry 1986). However the N effect alone cannot completely explain the positive effects of legumes on the succeeding crop yields. In the savannahs of West Africa, little information exists on the long-term effects of crop rotations, in particular the effects of N_2-fixing legume crops and short-term fallow, and selected fertilization treatments on soil fertility maintenance. The main hypothesis of this work is that soil fertility can be improved by an integrated management of N_2-fixing legumes or short-term (1 year) fallow in cropping systems, and use of low-cost mineral amendments (phosphate rock and dolomite), or organic amendments associated with low quantities of mineral fertilizers. The purpose of this research was to identify novel and socio-economic viable management options for soil fertility improvement and sustained crop yields in smallholder farmer's systems of West Africa.

Materials and Methods

The experiments were carried out over a period of 14 years (from 1993 to 2002) at the agronomic research station of Farakô-Ba (4° 20′ West, 11° 6′ North and 405 m altitude), located in the Guinean savannah zone of Burkina Faso. This agro-ecological zone has one rainy season per year, starting in May-June and ending in October. High variations of rainfall were observed both between years and within the years of experimentation (data not shown). In general, planting was done in June and harvesting in October.

Agronomic Experiment

A long-term experiment (1993–2007) was laid down in a 6 years old fallow field. The soil was an Ultisol of low fertility, representative of the study area. The main properties of the original soil at the start of the experiment were as follows: weakly acid, sandy soil with low clay and organic carbon contents. Available P (P-Bray I), Calcium (Ca), Magnesium (Mg) and exchangeable Potassium (K) were very low. As a consequence of low clay and organic C contents, ECEC was very low (Bationo and Mokwunye 1991a). Soil samples were taken from the top 20 cm depth in the experimental plots in 1993 (first year) and 9 years after (2001) for laboratory analysis to assess changes in soil properties. Soil pH was measured in 1 N KCl using a 2:1 solution to soil ratio and exchangeable acidity was measured as described by McLean (1982). Organic carbon was measured by the wet chemical digestion procedure of Walkley and Black (1934). Total nitrogen was determined by the Kjeldahl procedure (Bremner and Mulvaney 1982). Calcium and Mg were determined by atomic absorption spectrophotometry, while K and Na were determined using flame photometry. Effective Cation Exchange Capacity of the soil (ECEC) was calculated by the total exchange bases. Available phosphorous was determined using Bray I method (Fixen and Grove 1990).

A factorial of four cropping systems and eight fertilization treatments in split plot design with randomised block arrangement was used. Four cropping systems were used as first factor in the main plots: cotton-groundnut-sorghum and fallow-sorghum rotations were compared to mono cropping of sorghum, and a continuous groundnut was included. Each main plot was split into eight sub plots corresponding to the fertilization treatments, which were included as second factor. As the rate of applied mineral fertilizers was relatively low, the fertilization treatments included other nutrient sources, which are locally available. The fertilization treatments were: NPK fertilizer, NPK + Crop Residues, NPK + Dolomite, PK + Crop Residues, PK + Manure, PK + Compost, PK alone and control (without any fertilization). Randomised Fisher blocks arrangement was used. Fertilization treatments were also randomised in the sub plots. Recommended rates of mineral fertilizers were applied to crops using urea, complex NPK fertiliser, triple super phosphate and potassium chloride (KCl). Nitrogen, P, K and S nutrients were applied to groundnut at rates of 14, 10, 11 and 6 kg ha^{-1} respectively, 37, 10, 11 and 6 kg ha^{-1} on sorghum and 44, 15, 17 and 9 kg ha^{-1} on cotton. NPKS fertilisers were applied as complex NPK fertiliser at sowing on all crops. At 40 days after sowing, complementary doses of 23 and 30 kg N ha^{-1} were applied to sorghum and cotton, respectively, as urea. To the manure and compost containing treatments, 3 t ha^{-1} of air-dried cattle manure or compost were applied. Cattle manure contained 1.8%, 18.40%, 0.31% and 0.16% of N, C, P and K while compost supplied 1.0%, 15.5%, 0.17% and 0.11% of N, C, P and K. Annual applications of 1 t ha^{-1} of dolomite (249 kg Ca ha^{-1} and 114 kg Mg ha^{-1} of Ca and Mg respectively) were used.

Recommended improved varieties of groundnut (RMP-12), sorghum (Gnonfing) and Cotton (FK 290) and a planting density of 62,500 plants ha^{-1} was used for the three crops. Crop yields were measured at the end of each growing season.

Isotopic Experiment

In the year 2001, dry matter yields of sorghum were evaluated at 15, 30 and 50 days after sowing. Soil mineral nitrogen and N fertilizer recovery were also measured. Soil mineral nitrogen ($NH4^{+}$ + $NO3^{-}$) was measured during the first 2 months of the season. Composite soil samples were taken from the 20 cm level topsoil at sowing, 8, 15, 29 and 46 days after sowing in the sorghum plots. Mineral N was extracted with a 1 M KCl solution and measured by a colorimetric method, according to Keeney (1982).

Fertilizer nitrogen recovery by sorghum was studied by the ^{15}N isotopic method for all the four cropping systems but in only three sub-plot fertilization treatments (Control, NPK+Dolomite and PK+Manure). Fourteen kg N ha^{-1} were applied as urea with 5 atomic % ^{15}N excess at sowing in the isotope micro plots laid down in the experimental plots. Harvest of sorghum was made at physiological maturity.

Harvested plant samples were dried at 60°C during 72 h and ground. Total N and atomic ^{15}N % excess was determined by an elemental analyser coupled to an isotope ratio mass spectrometer at the IAEA Seibersdorf Laboratory.

Indirect method was used for N recovery studies. Labelled urea with ^{15}N was used as tracer for N recoveries assessment. Fertiliser N use efficiency (FUE) was calculated using the percentage of N derived from fertiliser (Nddf) and the total N applied by fertilization treatment (Eq. 2.1).

$$\text{NUE} = (\text{Ndff - kg ha}^{-1} / \text{N applied - kg ha}^{-1}) \times 100 \tag{2.1}$$

Statistical analysis has been done using SYSTAT (version 8.0) software. Factorial procedure was used with cropping systems (main plots) as main factor and fertilizer treatment (sub plots) as second factor.

Results and Discussion

Crops Yields

Sorghum, cotton and groundnut yields were affected by fertilization treatments and cropping season (Table 2.1). However cropping season (year effect) greatly affected crop yield responses to both fertilization and cropping system. High variations in rainfall distribution during the growing season were observed over the experimental period of 14 years (Data not shown). However significant correlations were not observed between crop yields and total rainfall. Crop yields were also not correlated with rainfall distribution pattern over years indicating that rainfall variations didn't explain crop yield responses. High crop yield productions after fallow and slow yield decrease during the next 5 years are sometimes observed in Ultisols of this ecological zone (Bado et al. 1997). Fauck et al. (1969) explained this phenomenon by soil biological activity evolution towards a new equilibrium leading to lower crop productivity.

Sorghum

Sorghum grain yields were affected by fertilization treatments, cropping system and cropping season (Table 2.1, Fig. 2.1). Significant interaction was not observed between cropping system (rotations) and fertilization treatments meaning that rotation effects were not affected by fertiliser applications and conversely.

Fertilizer applications significantly increased sorghum grain yields during the 9 years. This is likely due to the original low fertility of the soil (Pichot et al. 1981; Pieri 1989; Sédogo et al. 1991; Bationo and Mokwunye 1991b). Highest total N uptake were obtained when mineral NPK fertiliser was associated with dolomite or

Table 2.1 Sorghum, cotton and groundnut yields (kg ha^{-1}) as affected by fertilization and cropping systems (14 years)

	Sorghum	Cotton	Groundnut yields	
	Grainy yields	Yields	Grain	Haulms
Fertilization				
NPK + dolomite	1,125[a]	1,287[a]	852	2,688[a]
PK + manure	1,087[ab]	1,180[b]	825	2,595[ab]
NPK + crop residues	1,011[c]	1,087[c]	730	2,218[cd]
PK + compost	965[cd]	1,035[cd]	813	2,381[abc]
NPK fertilizer	932[de]	986[cde]	660	1,839[de]
PK + crop residues	778[f]	857[f]	703	1,992[de]
PK fertilizer	632[g]	773[g]	664	1,977[de]
Control	473[h]	656[h]	564	2,056[cde]
Cropping systems				
Cotton-Groundnut-Sorghum	1,021[a]	943	735	2,279[a]
Fallow-Sorghum	912[b]	Na	Na	Na
Sorghum-Sorghum	547[c]	Na	Na	Na
Groundnut-Groundnut	Na	Na	715	2,099[b]

Values affected by the same letter in the same column are not significantly different at $p<0.05$, according to Fisher's test
Na Not applicable

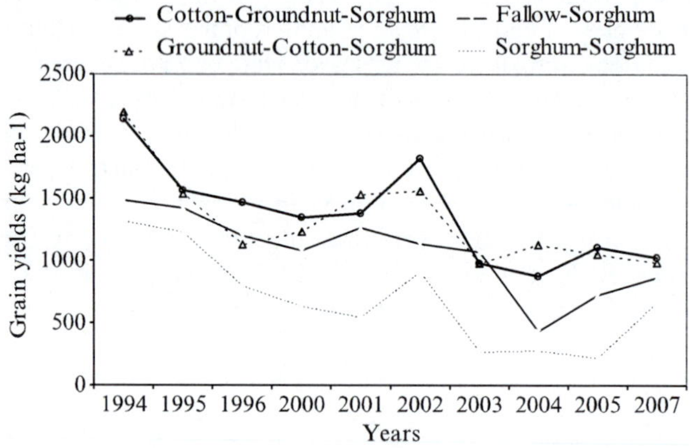

Fig. 2.1 Influences of crop rotations on succeeding sorghum grain yields during 13 years (1994–2007) of cultivation at Farakô-Ba. Data of years 1997 and 2006 were accidentally lost

when mineral PK fertiliser was associated with cattle manure (Table 2.2). Despite the lack of mineral N, the combined application of PK fertiliser with manure was as effective as NPK fertiliser, indicating that the N provided by manure could be as effective as the N supplied by NPK fertiliser.

With regard to cropping systems, sorghum produced highest grain yields when groundnut was included in the cropping system (cotton-groundnut-sorghum) while

Table 2.2 Total N uptake and fertilizer N use efficiencies (FNUE) of sorghum as affected by cropping systems and fertilizer after eight seasons of cultivation (1993–2001)

	Total N-uptake (kg ha^{-1})		NUE
Cropping systems	Total	Ndfr	%
Cotton-Groundnut-Sorghum	119[a]	71	51[a]
Fallow-Sorghum	104[abc]	56	32[c]
Sorghum-Sorghum	48[d]	0	19[d]
Fertilization			
NPK	78[c]	0	37
NPK + dolomite	98[b]	20	35
PK + manure	115[a]	37	38

Values affected by the same letter in the same column are not significantly different at $p<0.05$, according to Fisher's test

NFUE Nitrogen fertiliser use efficiency, *Ndff, Ndfs* N derived from fertilizer or soil respectively, *Ndfr* N differences between fertiliser or cropping system treatment and control (NPK or sorghum-sorghum rotation)

the lowest yields were obtained in mono cropping of sorghum (Table 2.1, Fig. 2.1). Fallow-sorghum rotation significantly increased sorghum grain yields compared to mono cropping of sorghum. Mean annual grain yields of sorghum were increased by 1.7 and 1.9 times when sorghum was rotated with fallow and groundnut respectively (Table 2.1). Beneficial effects of the inclusion of N_2-fixing legume crops on the production of succeeding non-fixing crops have been reported in many works (Peoples et al. 1995; Wani et al. 1995; Chalk 1998; Bationo and Ntare 2000). Peoples and Crasswell (1992) and Kouyaté et al. (2000) reported that legume-cereal rotations can increase cereal yields from 50% to 350%.

Cotton and Groundnut

Because cotton was included in only one cropping system (cotton-groundnut-sorghum), only its response to fertilization is discussed (Table 2.1). Fertilization influenced crop yield responses of cotton. Cotton yields were significantly increased by the combined application of mineral NPK fertiliser and dolomite compared to mineral NPK fertiliser alone. As found with sorghum, highest grain yields of cotton were obtained when NPK fertiliser was associated with dolomite, indicating that soil acidity (and aluminium toxicity) correction can significantly improve both sorghum and cotton yields (Berger et al. 1987; Bado et al. 1993). Annual mean grain yields were also high when PK fertilizer was applied with manure, meaning that manure supplied enough nitrogen for cotton as it was also found for sorghum.

Cropping systems and fertilization didn't affect groundnut grain yields (Table 2.1). However, the three factors (cropping systems, fertilization and cropping season) affected groundnut haulm yields. As already mentioned for sorghum and cotton,

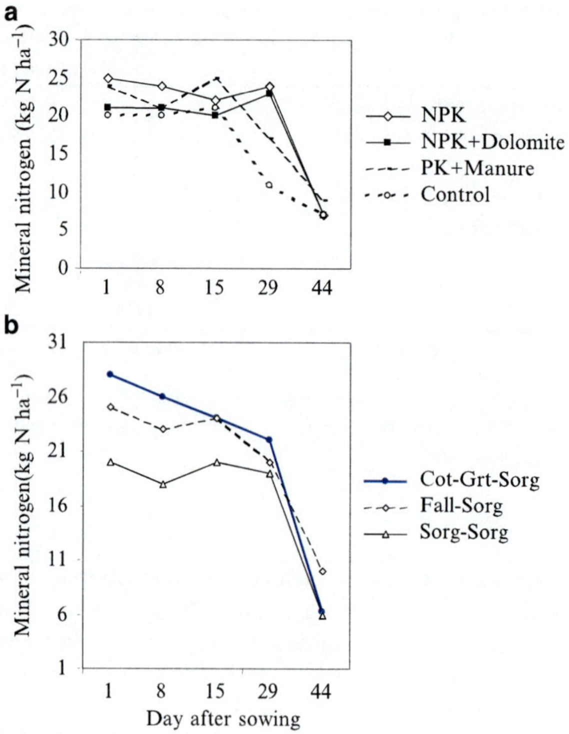

Fig. 2.2 Effects of (**a**) fertilizers and (**b**) crop rotations on soil mineral nitrogen during the first 44 days after sowing in 2001 (after 8 years of cultivation). *Cot* Cotton, *Grt* Groundnut, *Fall* Fallow, *Sorg* Sorghum

dolomite applications had significant effect on groundnut haulm yields and mean annual highest grain yields were obtained when NPK fertiliser was applied with dolomite. Annual mean haulm yields of groundnut with PK fertilizer associated with organic manure (2.6 t ha^{-1}) were similar to yields obtained when NPK fertilizer was applied with dolomite (2.7 t ha^{-1}).

Soil Mineral Nitrogen

During the rainy season of year 2001 (after 9 years of cultivation), soil mineral N decreased with time during the growing season (Fig. 2.2). The highest quantity of soil mineral N was found at sowing, decreasing with time first slowly up to 30 days after sowing and thereafter very rapidly. Very low quantity of mineral N was measured 44 days after sowing (DAS). Nitrogen application as NPK fertilizer or manure supplied most of the initial mineral soil N. Moreover in the control treatment it is likely that the first rains of the season stimulated the microbial activity leading to a high mineralization of soil organic matter resulting in an increase in mineral N. Decreasing soil mineral N over the season in all treatments was due to declining N mineralization rates and increased N uptake/demand by plants, and also probably to leaching and other N losses.

Cropping systems had significant effects on soil mineral N at sowing and interaction was not observed between the two factors ($P<0.05$). During the first 29 DAS, highest quantities of mineral N were observed in Cotton-Groundnut-Sorghum rotation while lowest quantities were observed in continuous sorghum. High mineral N in Cotton-Groundnut-Sorghum rotations could be a consequence of readily decomposable organic N provided by groundnut residues. Similar results on soil mineral N increasing when legumes are included in rotations were reported by others workers (Shumba 1990; Wani et al. 1995; Bationo and Ntare 2000). Despite the exportation of legumes fodders from the field, the remaining residues and the below ground parts of groundnut can contribute to improve soil organic matter and thereby soil mineral N (Chalk 1998). Compared to continuous sorghum, the good quality of groundnut residues probably supplied easily decomposable organic N resulting in increased levels of soil mineral N at the beginning of the next season. As reported by other researchers (Barrios et al. 1998; Bagayoko et al. 2000; Bationo and Ntare 2000; Bloem and Barnard 2001), soil mineral N is a good indicator to assess the N contributions of the precedent legumes to soil mineral N. More mineral N was found in the Fallow-Sorghum rotation than in continuous sorghum. This may likely be explained by the higher quantity of organic residues supplied by the fallow.

Nitrogen Recoveries

During the raining season of year 2001 (after 9 years of cultivation), total N uptake by succeeding sorghum and fertilizer N use efficiency (NUE) were significantly affected by cropping systems ($P<0.001$) and interactions were not observed between the two factors (Table 2.2). Compared to NPK fertilizer, highest N uptakes were observed with NPK + dolomite and PK + organic manure. Total N uptake increased from 48 kg N ha^{-1} in continuous sorghum to 104 and 119 kg N ha^{-1} when sorghum was rotated with Fallow and Cotton-Groundnut respectively. The effects of rotation and fertilizers on N yields increasing were also calculated using sorghum-sorghum rotation or NPK fertilizer as control treatments (Ndfr).

Fallow-sorghum and cotton-groundnut-sorghum rotations increased sorghum N yield from 56 to 71 kg ha^{-1} respectively (Table 2.2). Cowpea-sorghum and groundnut rotations increased N provided by soil from 59 to 73 kg ha^{-1} respectively. Compared to monoculture, sorghum took up 1.6 and 2.6 times more N from the fertilizer when it was rotated with Fallow and Cotton-Groundnut respectively, leading to high N fertilizer use efficiencies. The N fertilizer use efficiency values were 51% and 32% for the cotton-groundnut and fallow treatments, respectively, compared to only 19% for sorghum monoculture. The effects of legumes on increasing NUE have also been reported by others (Varvel and Peterson 1990; Mahadev-Pramanick and Pramanick 2000). Soils in the Fallow-Sorghum and Cotton-Groundnut-Sorghum treatments also supplied 2.3 and 2.5 times more N to sorghum respectively, in comparison to sorghum monoculture. These results clearly indicate that high total N uptake values when sorghum was rotated with fallow or cotton-groundnut were due to the increased nitrogen supply from sources, the fertilizer and the soil. As shown by data on soil mineral N, legume residues provide more organic N leading to high soil mineral N for the succeeding crop (Varvel and Peterson 1990).

Soil Properties

After 9 years of cultivation (1993–2001), cropping system effects on soil properties were examined with reference to the original soil. Soil organic carbon was significantly influenced by crop rotations and fertilization (Table 2.3). Highest quantities of soil organic carbon were observed when sorghum was rotated with fallow while lowest quantities were obtained for mono cropping of sorghum. The cotton-groundnut-sorghum rotation did not significantly increase soil organic carbon compared to mono cropping of sorghum. Only fallow-sorghum rotation maintained soil organic carbon at same level of the original soil. Within fertilizer's treatments, PK + manure applications increased soil organic carbon compared to NPK fertiliser alone. However, NPK + dolomite applications did not affect soil organic carbon compared to NPK fertilizer alone.

Soil total nitrogen was also affected by cropping systems and fertiliser applications (Table 2.3). All rotations decreased soil total N compared to original soil. Like organic C, highest quantities of total N were observed in soil when sorghum was rotated with fallow while lowest quantities were obtained in sorghum monoculture. But Cotton-Groundnut-Sorghum rotation did not significantly increase soil total N compared to mono cropping of sorghum. PK+Manure applications increased soil total N compared to NPK fertiliser.

Soil acidity parameters (pH, exchangeable aluminium and exchangeable acidity) and exchangeable bases were affected by crop cropping systems and fertilization. Except for Fallow-Sorghum rotation, all cropping systems increased aluminium saturation and decreased soil pH compared to original soil. But only soils of Fallow-Sorghum rotations maintained exchangeable acidity and base saturation at same levels like those of original soil. Compared to mineral NPK fertilisers alone, combined applications of NPK with manure or dolomite decreased exchangeable acidity and maintained soil pH and base saturation at same levels like those of original soil. On the contrary, NPK fertilizer alone tended to increase exchangeable acidity, lowering pH and base saturation.

Initial soil available P was very low, indicating that P was deficient in this soil and it is a limiting factor for crop production (Bationo and Mokwunye 1991b; Bado et al. 1997). Unlike organic C and total N, highest quantities of available P-Bray I were observed in soils of mono cropping of sorghum and lowest quantities of P-Bray I were observed in Fallow-Sorghum rotation. This can be explained by the higher crop yields and P uptake in the rotation treatments compared to mono cropping.

Table 2.3 Effects of fertilization and crop rotations on soil properties (0–20 cm layer) after 8 years (1993–2000) of cultivation

	pH water	pH KCl	Organic carbon (%)	P-BrayI mg P kg⁻¹	N-total mg N kg⁻¹	K^+ Cmol+ kg⁻¹	Ca^{++} Cmol+ kg⁻¹	Mg^{++} Cmol+ kg⁻¹	Al^{3+} Cmol+ kg⁻¹	AE Cmol+ kg⁻¹	ECEC Cmol+ kg⁻¹	Total bases (%)
Cropping systems												
Cotton-Ground-Sorg	6.3[b]	4.9[b]	0.39[b]	6.2[b]	373[c]	0.17[b]	0.87[c]	0.45	0.08[a]	0.16[a]	1.82	81[b]
Fallow-Sorghum	6.4[b]	5.3[a]	0.54[a]	4.1[c]	462[b]	0.22[a]	1.16[a]	0.49	0.01[b]	0.06[c]	2.02	93[a]
Sorghum-Sorghum	6.3[b]	4.9[b]	0.36[bc]	7.9[a]	372[c]	0.21[a]	0.87[c]	0.42	0.08[a]	0.15[a]	1.83	79[b]
Fertilization												
NPK	6.0[e]	4.5[d]	0.41[bc]	6.7[b]	386[b]	0.15[c]	0.61[d]	0.22[d]	0.15[a]	0.28[a]	1.56[c]	65[d]
NPK + dolomite	6.9[a]	5.4[a]	0.43[b]	4.9[c]	374[c]	0.14[c]	1.60[a]	0.81[a]	0.00[c]	0.01[d]	2.60[a]	99[a]
PK + manure	6.4[b]	5.4[a]	0.51[a]	10.3[a]	459[a]	0.35[a]	0.97[b]	0.46[b]	0.00[c]	0.04[c]	1.88[b]	96[b]
Control	6.2[c]	4.9[c]	0.38[d]	3.2[d]	308[e]	0.18[b]	0.72[c]	0.30[c]	0.04[b]	0.12[b]	1.46[d]	83[c]
Original soil	6.5[a]	5.1[a]	0.55[a]	2.7[d]	506[a]	0.02[c]	1.08[b]	0.46	0.01[b]	0.08[c]	1.82	96[a]

Values affected by the same letter in the same column are not significantly different at $p<0.05$, according to Fisher's test
EA Exchange acidity, *ECEC* Effective Cation Exchange Capacity

In the fertilization treatments highest quantities of available P-Bray I were found when manure was applied combined with PK fertiliser due to both P supplied by the PK fertilizer and manure and the positive effect of organic manure on P availability (Bationo and Mokwunye 1991a).

Conclusions

Crop yields and agricultural productivity can be improved in low input cropping systems by an integrated management of short-term fallow, N_2-fixing legume crops such as groundnut, and use of low-cost mineral amendments such as dolomite and organic amendments associated with low quantities of mineral fertilizers providing viable management options for soil fertility improvement in smallholder farmers' systems in the West African Guinean zone. With the lack of traditional long-term fallow system, annual fallow can be used to improve soil fertility. But cropping system that includes groundnut (Cotton-Groundnut-Sorghum rotation) is more effective than fallow-sorghum system. In Cotton-Groundnut-Sorghum rotation, cotton and sorghum can be efficiently fertilized by recommended doses of mineral NPK fertilizer. If available, 1 t ha^{-1} of dolomite can be applied on cotton. Apart from fallow-sorghum rotation, mineral PK fertilizer (without chemical N fertilizer) associated with 3 t ha^{-1} cattle manure is an effective and cost-effective option that should be applied on sorghum for sustainable management of soil fertility. However, the recommended dose of mineral NPK fertilizer (100 kg ha^{-1}) is appropriate for groundnut.

Acknowledgements This study has been funded by the "Institut de l'Environnement et de Recherche Agricole" (INERA), International Crops Research Institute for the Semi-Arid Tropics (ICRISAT), Tropical Soil Biology and Fertility Institute (TSBF) of the International Centre for Tropical Agriculture (CIAT) and International Atomic Energy Agency (IAEA). The authors are grateful to Dr. S. Koala, Director of ICRISAT, Sahelian Centre and Dr N. Sanginga, Director of TSBF Institute of CIAT for financial support. The authors are also grateful to Dr. F. Zapata, Project Officer, for revision of the draft manuscript and the staff of the FAO/IAEA Agriculture and Biotechnology Laboratory for providing ^{15}N labelled fertilizer and assisting in isotopic sample analysis under research contract No.BKF-10952 of the FAO/IAEA Co-ordinated Research Project on Tropical Acid Soils.

References

Bado BV, Dakyo D, N'Dayegamiye A, Cescas MP (1993) Effets de la dolomie sur la production et les propriétés chimiques d'un sol ferrallitique. Agrosol VI(2):22–24

Bado BV, Sedogo MP, Cescas MP, Lompo F, Bationo A (1997) Effet à long terme des fumures sur le sol et les rendements du maïs au Burkina Faso. Cah Agric 6(6):571–575

Bagayoko M, Buerkart A, Lung G, Bationo A, Römheld V (2000) Cereal/legume rotation effects on cereal growth in Sudano-Sahelian West Africa: soil mineral nitrogen, mycorrhizae and nematodes. Plant Soil 218:103–116

Barrios E, Kwesiga F, Buresh RJ, Sprent JJ, Coe R (1998) Relating pre-season soil nitrogen to maize yield in tree legume-maize rotations. Soil Sci Soc Am J 62:1604–1609

Bationo A, Mokwunye AU (1991a) Role of manure and crop residues in alleviating soil fertility constraints of crop production with special reference to the Sahelian and Sudanian zones of West Africa. Fert Res 29:117–125

Bationo A, Mokwunye AU (1991b) Alleviating soil fertility constraints to increased crop production in West Africa: the experience of the Sahel. In: Mokwunye A (ed) Alleviating soil fertility constraints to increased crop production in West Africa. Kluwer Academic Publishers, Dordrecht, pp 195–215

Bationo A, Ntare BR (2000) Rotation and nitrogen fertilizer effects on pearl millet, cowpea and groundnut yield and soil chemical properties in a sandy soil in the semi-arid tropics. West Afr J Agric Sci 134:277–284

Berger M, Belem PC, Dakouo D, Hien V (1987) Le maintien de la fertilité des sols dans l'Ouest du Burkina Faso et la nécessité de l'association agriculture-élévage. Cotton Fibre Trop XLII(3):201–207

Bloem A, Barnard RO (2001) Effect of annual legumes on soil nitrogen and on subsequent yield of maize and grain sorghum. South Afr J Plant Soil 18:56–61

Bremner JM, Mulvaney CS (1982) Nitrogen total. In: Page AL, Millet RH, Keeney DR (eds) Methods of soil analysis, vol 9, Agronomy Monograph. American Society of Agronomy, Madison, pp 595–622

Chalk PM (1998) Dynamics of biologically fixed N in legume-cereal rotations: a review. Aust J Res 49:303–316

Dommergues YR, Ganry F (1986) Biological nitrogen fixation and soil fertility maintenance. In: Uzo Mokwunye A, Vlek Paul PLG (eds) Management of nitrogen and phosphorous fertilizers in sub-Saharan Africa. Martinus Nijhoff Publishers, Dordrecht, pp 33–57

Fauck R, Moreaux CI, Thomann C (1969) Bilan de l'évolution des sols de Séfa (Casamance, Sénégal) après 15 années de culture continue. Agron Trop 24:213–301

Fixen PE, Grove JH (1990) Testing soil for phosphorus. In: Westerman RL (ed) Soil testing and plant analysis. Soil Science Society of America, Madison, pp 141–180

Keeney DR (1982) Nitrogen availability indices. In: Page AL, Millet RH, Keeney DR (eds) Methods of soil analysis, part II, vol 9, 2nd edn, Agronomy Monograph. ASA-SSSA, Madison, pp 711–730

Kouyaté Z, Kathrin-Franzluebbers, Juo-Anthony SR, Hossner-Lloyd R (2000) Tillage, crop residue, legume rotation and green manure effects on sorghum and millet yields in the semiarid tropics of Mali. Plant Soil 225:141–151

Mahadev-Pramanick MD, Pramanick M (2000) The effect of fertilizer and organic manure on rice-based cropping sequences in West Bengal. Trop Agric 77:111–113

McLean EO (1982) Soil pH and lime requirement. In: Page AL, Millet RH, Keeney DR (eds) Methods of soil analysis, part II, vol 9, 2nd edn, Agronomy Monograph. ASA-SSSA, Madison, pp 199–223

Peoples MB, Crasswell ET (1992) Biological nitrogen fixation: investment, expectation and actual contribution to agriculture. Plant Soil 141:13–39

Peoples MP, Herridge DF, Ladha JK (1995) Biological nitrogen fixation: an efficient source of nitrogen for sustainable agricultural production. Plant Soil 174:3–28

Pichot J, Sédogo MP, Poulain JF (1981) Évolution de la fertilité d'un sol ferrugineux tropical sous l'influence des fumures minérales et organiques. Agro Trop 36:122–133

Pieri C (1989) Fertilité des terres de savane. Bilan de 30 années de recherche et de développement agricoles au sud du Sahara. Agridoc-International, Paris

Sédogo MP, Bado BV, Hien V, Lompo F (1991) Utilisation efficace des engrais azotés pour une augmentation de la production vivrière: L'expérience du Burkina Faso. In: Mokwunye A (ed) Alleviating soil fertility constraints to increased crop production in West Africa. Kluwer Academic Publishers, Dordrecht, pp 115–123

Shumba EM (1990) Response of maize in rotation with cowpea to NPK fertilizer in a low rainfall area. Zimb J Agric Res 28:39–45

Steiner KG (1991) Overcoming soil fertility constraints to crop production in West Africa: impact of traditional and improved cropping systems on soil fertility. In: Mokwunye AU (ed) Alleviating soil fertility constraints to improve crop production in West Africa. Kluwer Academic Publishers, Dordrecht, pp 69–93

Varvel GE, Peterson TA (1990) Nitrogen fertilizer recovery by corn in monoculture and rotation systems. Agron J 82:935–938

Walkley A, Black JA (1934) An examination of the Detjareff method for determining soil organic matter and a proposed modification of the chromatic acid titration method. Soil Sci 37:29–38

Wani SP, Rupela OP, Lee KK (1995) Sustainable agriculture in the semi-arid tropics through biological nitrogen fixation in grain legumes. Plant Soil 174:29–49

Chapter 3
Crop and Soil Response to Tillage and Crop Residue Application in a Tropical Ferralsol in Sub-humid Western Kenya

J. Kihara, J. Mukalama, F.O. Ayuke, S. Njoroge, B. Waswa, J. Okeyo, S. Koala, and A. Bationo

Abstract Conservation agriculture (CA) offers an opportunity to reverse prevailing land degradation and consequent loss of productivity often occasioned by intensive soil tillage in cropping systems in most parts of sub-Saharan Africa (SSA). A long term experiment was established in Nyabeda Western Kenya in 2003 to evaluate the effect of tillage and crop residue application on maize and soybean yields, and on soil properties. The experiment was set up as a split-split-split plot design with four replicates and involved a factorial combination of tillage system (reduced and conventional tillage), cropping system (continuous cereal, soybean-maize rotation and intercropping), crop residue – maize stover – management (plus and minus crop residue) and nitrogen (N) application. Results showed that tillage influenced

J. Kihara (✉) • J. Mukalama • S. Njoroge • S. Koala
Tropical Soil Biology and Fertility Institute of International Centre for Tropical Agriculture (CIAT-TSBF), ICIPE Duduville Campus off Kasarani Road, P.O. Box 823-00621, Nairobi, Kenya
e-mail: J.Kihara@cgiar.org

F.O. Ayuke
Department of Land Resource Management and Agricultural Technology, The University of Nairobi, Nairobi, Kenya

B. Waswa
Tropical Soil Biology and Fertility Institute of International Centre for Tropical Agriculture (CIAT-TSBF), ICIPE Duduville Campus off Kasarani Road, P.O. Box 823-00621, Nairobi, Kenya

Centre for Development Research, University of Bonn, Bonn, Germany

J. Okeyo
Tropical Soil Biology and Fertility (TSBF), Institute of CIAT, c/o ICRAF, Nairobi, Kenya

University of Wyoming, Laramie, WY, USA

A. Bationo
Alliance for a Green Revolution in Africa (AGRA), Accra, Ghana

A. Bationo et al. (eds.), *Lessons learned from Long-term Soil Fertility Management Experiments in Africa*, DOI 10.1007/978-94-007-2938-4_3,

performance of maize although significant tillage effects were observed in only 5 out of the 15 seasons analyzed. Overall average maize grain yields were 2.9 ton ha^{-1} in reduced tillage and 3.6 ton ha^{-1} in conventional tillage systems. Application of crop residue increased seasonal maize grain yield in reduced tillage (340 kg ha^{-1}) and in conventional tillage (240 kg ha^{-1}), but the only significant crop residue (CR) effect was observed in season 10. Differences in maize yields between the two systems were attributed to phosphorus availability as it was demonstrated that application of crop residue in the reduced tillage resulted in better availability of P than without crop residue application.

Under the rotation system, significant tillage effects were observed in 6 of the 15 seasons with greater maize yield in conventional than in reduced tillage system. Soybean yields under reduced tillage were comparable to those from conventional tillage with the good performance of soybean in reduced tillage being related to the effect of its canopy on soil evaporation, and or to changes in microbial diversity and soil structure. For both the conventional and reduced tillage systems, legume benefits on succeeding maize were observed with similar maize yields being observed between maize monocropping and maize rotated with soybean. Reduced tillage improved soil aggregation with greater aggregate mean weight diameters being observed in this system than in conventional tillage. Tests for biological activity showed that the application of crop residue increased termite abundance in both reduced and conventional tillage systems. The results from this study indicate the importance of long term trials in better understanding the beneficial effects of conservation agriculture on soil productivity.

Keywords Reduced tillage • Rotation • Soybean • Soil aggregation • Soil organisms

Introduction

Intensive soil tillage in cropping systems has been blamed for the prevailing land degradation and consequent loss of productivity in many parts of sub-Saharan Africa (Fowler and Rockstrom 2001; Derpsch 2008). Consequently, this has formed a basis for the introduction of conservation agriculture (CA) practices with minimum soil disturbance (McCarty and Meisinger 1997; Wander and Yang 2000). Benefits associated with CA include include erosion control, increased infiltration, soil moisture conservation, reduced costs of fuel consumption (Benites 2008; Derpsch 2008) and drudgery (Landers 2008), improved soil ecology and nutrient cycling (Blank 2008), food security, increased incomes and a variety of environmental services (reduced CO_2 emissions, less flood damage, cleaner surface waters, (Landers 2008)). Studies have identified minimal soil disturbance, surface residue retention and crop rotation (Erenstein et al. 2008) and restricted in-field traffic (Benites 2008) as important principles of CA. Examples of practices that obey the CA concept include mulch-based no-till systems (known variably as zero-tillage or direct seeding) with soil disturbance restricted to seed sowing (Erenstein et al. 2008); and conservation tillage involving some form of soil disturbance, e.g., strip-tillage, ripping and subsoiling, ridging, and varied locally-adapted reduced tillage practices (Erenstein et al. 2008).

Like in many other countries in sub-Saharan Africa, conservation tillage had not been widely adopted in Kenya (Rockstrom et al. 2003) but there is clear momentum for this presently. While a few farmers have adopted practices involving the use of mulch, ripping and sub-soiling without inverting soil (Gitonga et al. 2008), the majority have stuck to the conventional practice of seasonal and mid-season tillage operations. In eastern Africa, this has mainly been attributed to the low crop productivity under conservation tillage compared to the conventional tillage systems commonly practiced by farmers (Taa et al. 2004), and there is the tendency to stick to 'custom' practices. The low crop productivity is occasioned by the fact that positive effects on crop productivity resulting from the effects of improved soil-plant moisture and nutrient relations in reduced tillage generally become evident only after long periods of applying the technique (Ghuman and Sur 2001; Malhi et al. 2006). Long-term trials with sufficient replication in time therefore have the ability of providing a better picture of the benefits of conservation tillage than short-term studies. Such long-term evaluations/experimentations are rare due to the lack of financial resources required for their maintenance. In addition, conservation agriculture studies in Kenya have traditionally concentrated on the drier zones where soil moisture is often limited and soil erosion is prevalent (Biamah 2005). In line with this, the Tropical Soil Biology and Fertility institute of CIAT (TSBF/CIAT) initiated a form of conservation tillage referred here as reduced tillage (RT). This practice involves the application of crop residues and the use of hand-hoes for weeding restricted to scratching the top 0–3 cm soil, only in the parts with weeds. In such a system, the source of mulch is often crop residue from the previous season, mainly cereals. Significant contributions to the quantity and quality of residue mulch can however be made from legumes in association with cereal. The inclusion of legumes in such a cropping system has been shown to have both short-term benefits (increased water and nutrient use efficiency, yields and economic returns) and long-term effects (increased N-supplying power of the soil, microbial diversity and carbon sequestration (Gan and Goddard 2008)).

Among the benefits of RT is the enhancement of processes such as soil fauna and microbial activities, both of which improve soil structure through the formation and stabilization of soil aggregates (Craswell and Lefroy 2001; Feller et al. 2001). The aggregates formed further improve the fertility status of soil through reduced SOM losses (Feller et al. 2001). Soil fauna and soil aggregation are ultimately coupled and long-term trials provide an opportunity to study management influences on fauna and soil aggregation over multiple periods.

The greatest challenge in reduced tillage systems is perhaps the availability of sufficient crop residues for use as mulch at the small-holder level. This has been identified as a common problem in many regions (Erenstein 2003; Tursunov 2009) including sub-Saharan Africa (Bationo et al. 2007; Fowler and

Rockstrom 2001). This affects the success of RT as recommendations suggest that at least 30% of the soil surface should be covered with crop residues at the time of planting for the expected effects of CA to be achieved (Fowler and Rockstrom 2001). However, the effect of the mulch in the conservation practice is further influenced by its quality and quantity (Scopel et al. 1998) as well as by the rate of disappearance of the residue through processes such as comminution by termites.

In this chapter, we evaluate the performance of different tillage and crop residue systems in two major cropping systems within a long-term experiment established in a sub-humid zone in western Kenya in 2003.

Materials and Methods

Study Site

The study was conducted in Nyabeda (located in a sub-humid site in western Kenya). The area is characterized by two rainy seasons: a long rainy season between March and August and short rainy season between September and February. Maize is the main staple crop and it is normally grown either as a monocrop or in association with legumes, mainly common beans and groundnuts. Adoption of soybean which is treated as a cash crop is also taking place in the region. Smallholder settlements predominates this area, with land sizes ranging from 0.3 to 3 ha per household. The dominant soil type in this area is Ferralsol (Table 3.1).

Table 3.1 Location, climatic and soil characteristics of the study site

Parameter	Nyabeda
Agro-ecological zone	Lower midland 2
Coordinates	0° 07′ N; 34° 24′ E
Altitude (m.a.s.l.)	1,420
Total annual rainfall (mm)	1,200–1,600
Daily temperatures (°C)	Mean = 23.2 (±1.52); min = 14; max = 31
Soil type	Ferralsol
Sand:silt:clay ratio	15:21:64
pH (water)	5.08 (±0.27)
Extractable K (me 100 g^{-1})	0.10 (±0.04)
P (mg P kg^{-1})	2.99 (±2.09)
Ca (cmolc kg^{-1})	4.69 (±0.33)
Mg (cmolc kg^{-1})	1.68 (±0.13)
Total SOC (%)	1.35 (±0.06)
Total nitrogen (%)	0.15 (±0.02)

Adapted from Kihara (2009)

Table 3.2 Treatments combining tillage practice, crop residue and cropping system

Treat. no.	Tillage method	Crop residue	Cropping system
1	Reduced tillage	Minus CR	Legume-cereal rotation
2	Reduced tillage	Minus CR	Legume-cereal intercropping
3	Reduced tillage	Minus CR	Continuous cereal
4	Reduced tillage	Plus CR	Legume-cereal rotation
5	Reduced tillage	Plus CR	Legume-cereal intercropping
6	Reduced tillage	Plus CR	Continuous cereal
7	Conventional tillage	Minus CR	Legume-cereal rotation
8	Conventional tillage	Minus CR	Legume-cereal intercropping
9	Conventional tillage	Minus CR	Continuous cereal
10	Conventional tillage	Plus CR	Legume-cereal rotation
11	Conventional tillage	Plus CR	Legume-cereal intercropping
12	Conventional tillage	Plus CR	Continuous cereal

Crop residue (maize stover) was applied at 2 t ha^{-1}

Experimental Design and Crop Management

The experiment was set up in 2003 as a split-split-split plot design with four replicates and involved a factorial combination of tillage system (reduced and conventional tillage), cropping system (continuous cereal, soybean-maize rotation and intercropping), crop residue – maize stover – management (plus and minus crop residue) and nitrogen (N) application (12 treatments; Table 3.2). For continuous maize, treatments were split to accommodate an N response, using 0, 30, 60 and 90 kg N ha^{-1}. Rotation plots were split into four to accommodate 0 and 60 kg N ha^{-1} each for legume and cereal crops. Since all treatments received phosphorus (P), additional rotation treatments without P were incorporated to enable investigation of P effects. All intercropping treatments received 60 kg P ha^{-1}, with no N applied.

Maize was planted at an intra row spacing of 0.25 m and an inter row spacing of 0.75 m with two seeds per planting hole and later thinned to one plant per hill (53,000 hills ha^{-1}). Soybean was planted at an intra row spacing of 0.05 m and an inter row spacing of 0.75 m (266,000 seeds ha^{-1}). A basal application of K in the form of muriate of potash was applied to all plots at 60 kg ha^{-1} while P (from triple super phosphate – TSP) and N (from urea) were applied at the specified rates every season. N was split-applied with $^{1}/_{3}$ applied at planting and $^{2}/_{3}$ at knee height (5 weeks after planting). Fertilizers were applied through the hill placement method.

At the onset of the experiment, land preparation was done uniformly across all plots by hand-ploughing to a 15 cm depth. Subsequent tilling operations involved hand-hoeing up to about 10 cm depth in conventional tillage, and surface scratching to about 3 cm depth only in portions with weeds in reduced tillage. There were two weeding operations and one hand pulling operation under continuous cereal and intercropping for each of the seasons. For rotation, while the part under maize was weeded in the same way as the continuous cereal plots, the part under soybean was weeded once and hand-pulled once every season. Crop residue (maize stover) was applied seasonally at 2 t ha^{-1} before planting, while fertilizers were hill-placed at the time of sowing.

Maize (*Zea mays L.*) varieties planted varied from season to season and included Hybrid 513, Hybrid 502 Imidazolinone-Resistant (IR) maize, and hybrid 403. For the legumes, common bean (*Phaseolus vulgaris L.*) was grown in season one and soybean (*Glycine max* (L.) Merr) was grown in the subsequent seasons. Legume residues were left on the respective plots. Prior to establishment of the experiment, the farm had been under a native vegetation of grasses and shrubs.

As in many long-term experiments, changes were made as new research questions came up. For example in 2005, to investigate the effect of residue loss on yields, rotation plots that received 60 kg N ha^{-1} were split into two with one part where fauna was excluded using insecticides and the other part where no insecticide was applied. The plots were treated with Endocoton, with endosulfan 35% EC as the active ingredient, at a rate of 450 g a.i/ha (approx. 0.9 l ha^{-1}.) to eliminate earthworms and Dursban with chlorpyrifos 44.6% w/w as the active ingredient at a rate of 400 g a.i/ha (approx. 0.8 l/ha.) to control termites (Hurisso 2007). Also, in 2006, an annex experiment was implemented to investigate effect of ripping and subsoiling, aspects that were felt important for success of reduced tillage systems.

Soil Characterisation

At the onset of the experiment, soil samples were obtained from 0 to 15 cm depth, bulked per replicate and analyzed for total C and N, available P and K, exchange Ca and Mg, texture and pH. Soil total N was based on wet oxidation using Kjeldahl digestion, while soil total C was extracted using acidified dichromate (Anderson and Ingram 1993). Soil pH was determined in water, while exchangeable Ca and Mg were extracted using 1 N KCl extractant and determined using a spectrophotometer. Analyses for extractable inorganic phosphorus and exchangeable potassium were based on a modified Olsen extractant, 0.5 M $NaHCO_3$ + 0.01 M EDTA, pH 8.5, (ICRAF 1995; Okalebo et al. 2002). Soil texture was determined using a Bouyoucous hydrometer (Hydrometer method) (Gee and Bauder 1986).

Maize and Soybean Yield Determination

Maize was harvested from the plots at maturity leaving two border plants (0.25 m spacing) on both ends of the row and one row (0.75 m spacing) from the other ends to eliminate edge effects. Cobs were separated from the stover and the fresh weight of each determined. After air drying the sub-sample, grains were separated from the cobs, oven-dried at 60°C for 48 h and their separate dry weights determined. Similarly, legume grain and biomass, harvested from equal plots as maize, were measured on an oven-dry basis.

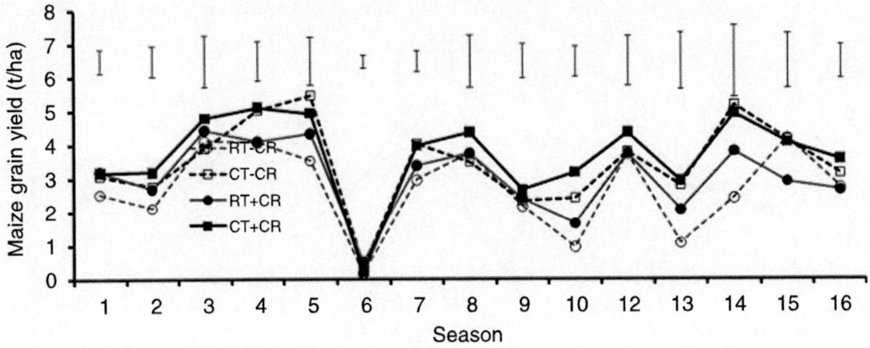

Fig. 3.1 Maize grain yield in continuous maize system in Nyabeda, western Kenya, all plots was applied with chemical N and P each at 60 kg ha^{-1}, *error bars* are 2 standard error of difference of means

Data Analysis and Presentation

All yield data was analysed using Genstat Discovery (seventh edition) statistical software for Windows. Since the purpose was to assess the effect of tillage and crop residue application, the data was analyzed for each season and cropping system separately. Soil aggregation and phosphorus data were analysed in SAS. Mixed model procedure was used for analysis of variance between treatments allowing for analysis of random effects of the replicate, tillage practice and applied crop residue in this split-split-split plot experiment. Least square means were compared directly in SAS using PDIFF. Soil microbial data was analysed using SPSS for windows.

Results and Discussion

Agronomic Performance

Tillage influenced performance of maize although significant tillage effects were observed in 5 out of the 15 seasons analyzed (seasons 4, 5, 7, 10 and 14). Overall, average maize grain yields were 2.9 ton ha^{-1} in reduced tillage and 3.6 ton ha^{-1} in conventional tillage systems (Fig. 3.1). Application of crop residue increased seasonal maize grain yield in reduced tillage (340 kg ha^{-1}) and in conventional tillage (240 kg ha^{-1}), but the only significant CR effect was observed in season 10. A similar effect was observed with stover yields (data not shown). These yield increases following crop residue application are more than double the income that a farmer would stand to gain presently by trading their CR for cash in Western Kenya. Separation of means showed yield under reduced tillage with crop residue application was in most cases similar to conventionally tilled treatments. Although it is still

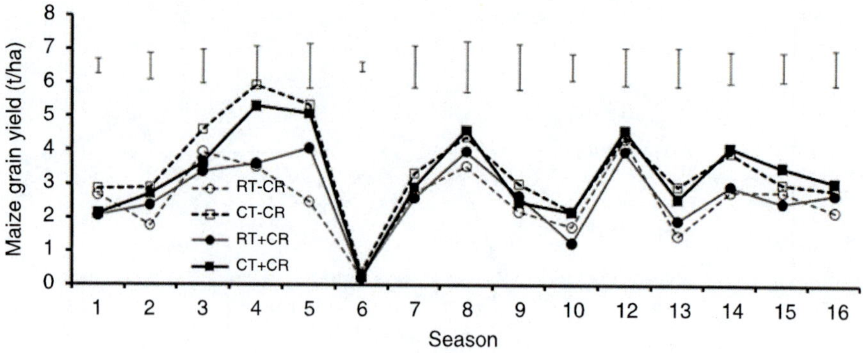

Fig. 3.2 Maize grain yield in maize-legume rotation system, No chemical N was added, all plot had P at 60 kg ha^{-1}, *error bars* are 2 standard error of difference of means

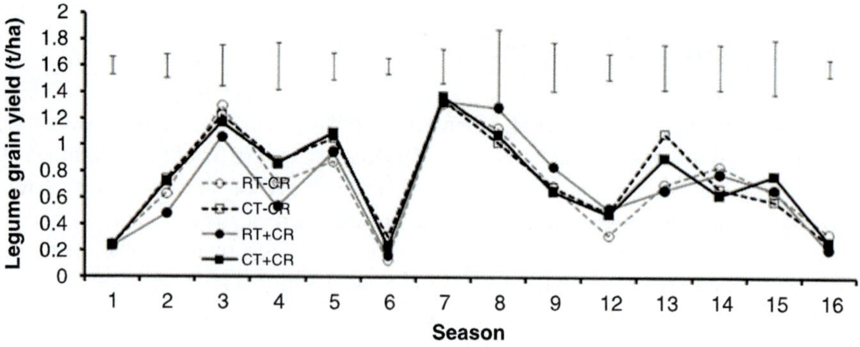

Fig. 3.3 Legume grain yield in maize-legume rotation system, No chemical N was added, all plot had P at 60 kg ha^{-1}, *error bars* are 2 standard error of difference of means

not clear what level of crop residue (maize stover) can support sufficient soil moisture and microclimate in the reduced tillage system in tropical Africa, the 2 t ha^{-1} used in this long-term experiment may be insufficient to boost productivity to match that of conventional tillage immediately reduced tillage is adopted. However, this is the amount that farmers are most likely to access and thus alternative soil cover such as green manure cover crops should be explored to supplement the limited crop residue available for mulch.

Legumes in cropping systems alter soil moisture and nutrient dynamics through litterfall, N fixation and other rhizosphere processes. Under rotation system in our long-term experiment, significant tillage effects were observed in 6 of the 15 seasons (seasons 2, 4, 5, 10, 13 and 14) with greater yield in conventional than in reduced tillage system (Fig. 3.2). Application of CR depressed maize yields notably in the conventional tillage system consistently up to the fifth season (because no chemical N was applied). The lack of yield depression following several rotation cycles (season 6 onwards) could imply improvement in soil health in this system.

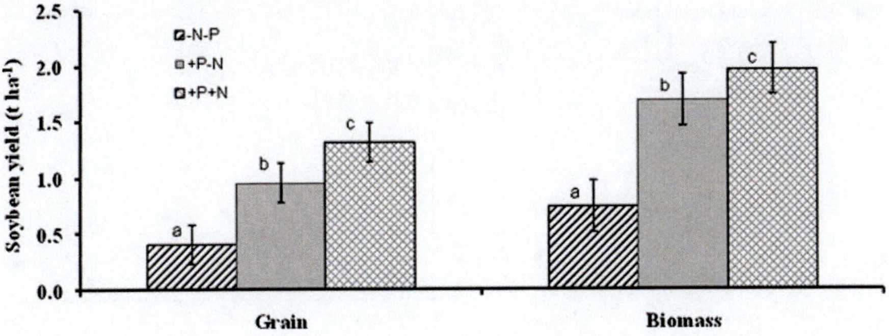

Fig. 3.4 Effect of phosphorus and nitrogen on soybean yield in soybean-maize rotation system over the first eight seasons in Nyabeda, western Kenya; *bars* are confidence limits; *different letters* indicate significant differences; legume yield did not include first season

Depressive effect of low quality residue application such as maize stover is often associated with N immobilization by soil microbes. In such systems, some chemical N may be required during those initial seasons.

Significant effect of tillage on soybean yield was observed only in two of the driest seasons (seasons 2 and 6) and in season 13. Soybean yields under reduced tillage were thus comparable to those from conventional tillage (Fig. 3.3), and soybean is a good candidate crop for conservation tillage systems under normal rainfall seasons. The good performance of soybean in reduced tillage could be due to the effect of its canopy that reduced soil evaporation. It could also be linked to changes in microbial diversity and soil structure, relative to tilled/disturbed system. Kihara (2009) quantified the amount of nitrogen fixed in the maize-legume systems as 53–82 kg N ha^{-1} although about 60% of this was in form of harvested soybean grain. The N fixation study also showed more NDfA in reduced compared to conventional tillage system. Thus soybean production in western Kenya is better done within reduced tillage.

Investigation of legume benefits on succeeding maize showed for both conventional and reduced tillage system. Similar maize yields were observed between maize monocropped and maize rotated with soybean even though no chemical fertilizer N was added in the rotation system. Rotational benefits of legumes to succeeding cereal have been attributed to soil quality aspects such as fixed N and solubilization of unavailable nutrients through root exudates, biologically as well as through improvement in soil physical properties.

N and P Fertilization

Nitrogen and phosphorus are crucial for crop production in western Kenya. High response to phosphorus fertilization (Fig. 3.4) is related to the low available P in these soils, also characterized by high P adsorption (Gachengo et al. 1999). Applying P alone was better than no P at all and treatments with only phosphorus added (+P-N) had more than double the yield from non-fertilizer treatments (–P-N). Legumes require P for effective root development and nitrogen fixation.

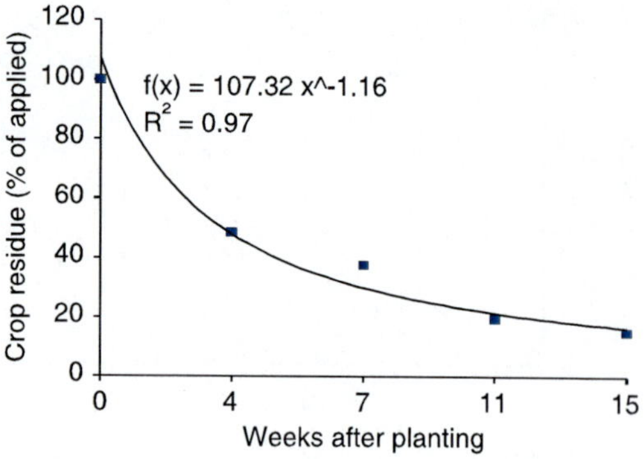

Fig. 3.5 Rate of crop residue disappearance observed in Nyabeda, western Kenya during long rains cropping season in 2007

Residue and Fauna Experiments

The rate of CR disappearance was assessed during the 10th cropping season and the results indicated that up to 40% of the residue is lost within the first month of application, attributed mainly to comminution by termites (Fig. 3.5). The reduced rate of residue disappearance with time is attributed to reduced quality of the remaining residue. Thus high presence of termites in the study site, including the *Pseudacanthotermes* and *Synacanthotermes* (Ndabamenye 2006) of the fungus-growing Macrotermitinae subfamily, family Termitidae (Krishna 1970) results in fast residue loss. In a related residue loss study conducted more recently in the same experiment, Terano (2010) estimated the applied 2 ton ha^{-1} maize stover to constitute 25% of surface residue cover in reduced tillage system but this reduced to about 17% within 1 week and only about 1% was under cover from 9 weeks after planting onwards. Kihara (2009), based on a litterbag study, attributed over 90% of residue loss in the first 1.5 months after application to comminution by macro and mesofauna mainly termites. This calls for the need to use other residue sources that are not attacked by termites. Leguminous species could have an appropriate niche within conservation tillage systems in western Kenya if these are introduced before maturity of the maize crop. Although legumes, especially herbaceous ones have been tried previously without much success, CA presents an opportunity to test and re-introduce them.

An assessment of termite abundance in the top 0–15 cm soil depth during a growing season in these plots showed extremely low number of termites in the plots applied with insecticide compared to plots without insecticide (average of 8 vs 266 termites per m^{-2}; (Terano 2010)). The agronomic results showed that plots where macrofauna were excluded had higher yields than plots where macrofauna were present (Fig. 3.6). These results were surprising as soil macro-fauna were expected to improve soil structure and result in better yields. Although the increased yields was suspected as being the influence of water harvesting through confinement by microplot metal sheets during the first season (Hurisso 2007), similar results were achieved during consecutive seasons even though metal sheets confining rainwater had been removed. It is thus unclear what specific influences the applied chemicals had to positively influence crop performance. However, although the surface cover

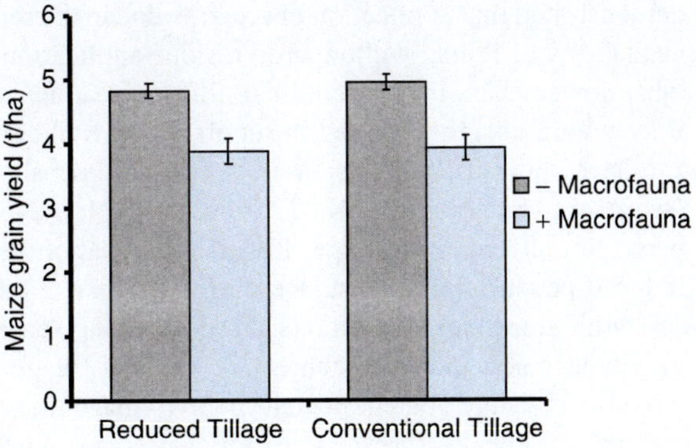

Fig. 3.6 Effect of macro-fauna on maize productivity in a maize-soybean rotation system in Nyabeda, western Kenya (data used is of 6 seasons between 2006 and 2010, treatments received N, P and K each at 60 kg ha^{-1}, chemicals were applied seasonally)

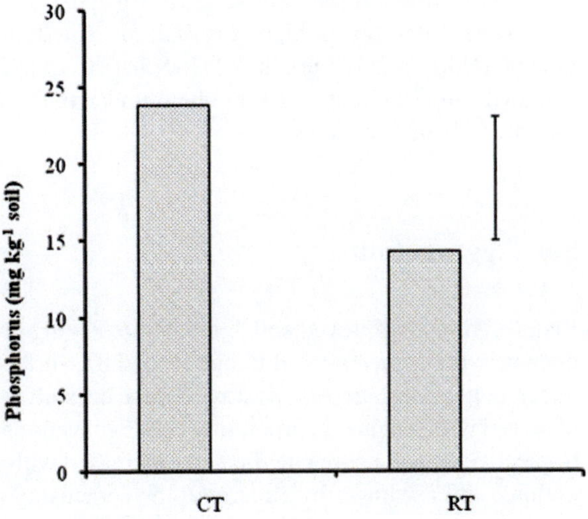

Fig. 3.7 Available soil phosphorus measured in the 0–15 cm soil depth in Nyabeda, western Kenya in 2007 (*RT* reduced tillage, *CT* conventional tillage, *CR* crop residue, *error bars* are 2× standard error (Adapted from Kihara 2009))

by residue was not so different between plots with and those without insecticide, the quantity of undecomposed residue was high in plots applied with insecticide (Terano 2010), an indication that soil moisture could play a part. Carbon and nitrogen was however by 5–14% more in insecticide treated plots (Terano 2010) which may explain some of the yield increases.

Soil Chemical Tests

The difference in yield between reduced tillage and conventional tillage in this site was attributed to phosphorus. For example, there was a mean of 24 mg P/kg of soil in conventional tillage system compared to 15 mg P/kg of soil under the reduced tillage system (Fig. 3.7) and data was consistent in 2007 and 2008. It was also demonstrated that application of crop residue in the reduced tillage resulted in better availability of P than without crop residue application although this was still somewhat lower than in conventional tillage treatments (Kihara 2009). Significant P losses are usually reported in runoff water. Soil compaction studies in our study site have showed slight increases in soil bulk density and penetrometer resistance (seasonal average of 298 vs 227 lb/in.^{2} at the top 0–5 cm) in reduced tillage compared to conventional tillage. Residue application in reduced tillage significantly reduced penetrometer resistance during the first 2 weeks of planting from 201 to 157 with some positive effects throughout the season (Terano 2010). Similarly, gravimetric soil moisture content in reduced tillage plus crop residue resembled more the moisture content in conventional tillage plots while reduced tillage without residue was always lowest for the top 0–30 cm soil depth (Terano 2010). Results of soil N, pH and Carbon at 0–15 cm depth showed no differences between treatments as at 2007. However, and consistently for the top 0–15 cm soil, reduced tillage with crop residue had the highest carbon concentration in 2005 (Ndabamenye 2006), in 2006 (Hurisso 2007) and in 2007 (Kihara 2009). Terano (2010) reported only a marginal effect of crop residue where under reduced tillage, residue application increased N and C contents.

Soil Aggregation

Reduced tillage improved soil aggregation, having greater aggregate mean weight diameters than conventional tillage (Fig. 3.8). Surface application of crop residue under reduced tillage had positive effect on soil aggregation but no effect was observed with residue incorporation under conventional tillage. As expected in this tropical Ferralsol, aggregate diameter increased with soil depth. Among the tested options, and as shown by Kihara (2009), practising reduced tillage (with residue

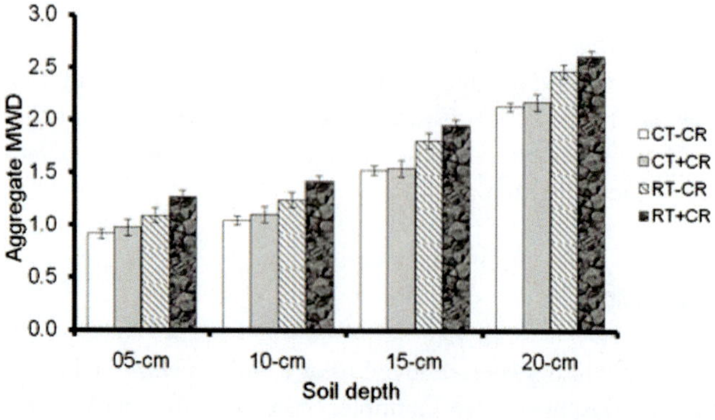

Fig. 3.8 Effect of crop residue application on soil aggregation (*MWD* mean weight diameter in mm) in Nyabeda, western Kenya after ten seasons of treatment application

application) is the best means to improve soil aggregation in this soils. Similar results were reported for the same experimental site by Ndabamenye (2006), using data collected in 2005, during the 7th cropping season (large macro-aggregates in reduced tillage were significantly greater than in conventional tillage). During the 9th season after establishment, Ayuke (2010) compared soil aggregation for this experiment with a nearby fallow and observed significantly greater soil aggregation in the fallow than the arable field, and this was consistent with the carbon content of those soils, an indication that soils with more carbon are likely to be better aggregated. An analysis of carbon concentration in the macroaggregates showed increased carbon on the macroaggregates obtained from reduced tillage plus crop residue compared to those from conventional tillage system and reduced tillage without residue application (Hurisso 2007). The benefits of soil aggregation in the reduced tillage system has not yet fully translated into increased yields and this is partly due to the fact that no ripping or subsoiling was done at the establishment of the trials. It has been suggested that subsoiling is necessary at the onset since the soils have been subjected to years of degradation with likely hardpans (Okoba, personal communication).

Soil Biological Activity

As expected, application of crop residue increased termite abundance in both reduced and conventional tillage systems. At the top 0–15 cm soil depth, Terano (2010) observed the number of termites to more than double between plots without residue and those with residue in most of the instances. A similar result was reported for long rains 2006 season by Hurisso (2007) where also other soil fauna increased in abundance (e.g., from 8 to 20 m^{-2}) with residue application under reduced tillage. The effect of crop residue on termite abundance extended even at the 15–30 cm soil depth where greater numbers of termites were observed in plots with residue addition (Terano 2010).

In another study where all treatments were supplied with crop residues, Ayuke et al. (2011) found no difference in earthworm and termite taxonomic richness, abundance and biomass among the treatments (Table 3.3), contrary to their hypothesis. They attributed this to the wide feeding and foraging habits of most of the species sampled. Very few typical soil feeders (group IV) were found. Rather, a large proportion of the termites sampled in this study belong to feeding group II and to a smaller extent to feeding group I and III (Ayuke et al. 2011). These groups are largely foragers of wood, litter, soil, dung/manure and grass. They were probably scavenging for food all over the plots and, hence, were sampled uniformly across treatments (Ayuke et al. 2011).

Earthworms sampled in this study site were largely comprised of the epigeic, to a small extent the endogeic and none of the anecic taxa. Epigeic earthworms, which need organic inputs as mulch, and endogeic species which are capable of utilizing low-quality organic residues as food, would have been favoured by the addition of

Table 3.3 Mean earthworm richness, abundance, and biomass under different cropping systems

Treatment	Richness	Abundance (No m^{-2})	Biomass (g m^{-2})
Earthworms			
MM + CT	1.33	43	3.93
MM + R	1.33	80	0.76
SM + CT	1.00	53	1.62
SM + RT	0.33	11	0.17
***p*-value**	**0.771**	**0.736**	**0.577**
Termites			
MM+CT	0.33	16	0.01
MM+RT	0.00	0	0.00
SM+CT	0.33	32	0.02
SM+RT	0.67	208	1.14
***p*-value**	**0.793**	**0.248**	**0.442**

All treatments were applied with crop residue. Based on monolith sampling done 8 weeks after planting (June 2007). One soil monolith of 25×25×30 cm depth was randomly taken in each replicate plot (n=3). The soil was separated into two depth layers (0–15 and 15–30 cm) and earthworms and termites were collected by hand-sorting on plastic trays. Earthworms and termites were identified at the Department of Invertebrate Zoology of the National Museums of Kenya (NMK), Nairobi, Kenya, using appropriate keys and reference specimens in the NMK collections. Adult earthworms and termites were identified to species or where this proved difficult, to numbered morphospecies. Unidentified earthworm specimens were sent to the Hungarian Natural History Museum (HNHM), Budapest, Hungary, for further analysis. Voucher specimens are being held at the NMK, Nairobi and HNHM, Budapest. Taxonomic richness (*S*) was calculated as the number of taxa per monolith. Due to non-homogeneity of variances in the earthworm and termite data (richness and abundance), these were square root transformed $(x+0.5)^{1/2}$

MM Maize-maize system, *SM* Soybean-maize rotation system, *CT* Conventional tillage, *RT* Reduced tillage

organic resources as used in our study site. Anecic earthworms were lacking across all treatments (Ayuke et al. 2011). Anecic earthworms tend to be few in tropical soils in general, due to low litter availability as a result of higher decomposition rates at higher temperatures (Lavelle et al. (1999)), and in arable fields in particular because they tend to be more sensitive to disturbance and tillage (Curry 2004). It can be concluded that CR is the main factor affecting soil fauna while tillage and rotation system have little effect.

Besides the effects on soil meso- and macrofauna, molecular investigations have shown that reducing tillage profoundly changes the soil microbial communities as shown in a study conducted by Kihara (2009; Fig. 3.9). In the study, the lowest diversity indices were observed in a maize-maize rotation system although this was in most cases not significantly different from maize-legume rotation system. Unlike with conventional tillage, crop residue application under reduced tillage resulted in a somewhat different soil bacteria community (Kihara 2009). The resulting microbial communities due to reduced tillage can have great impacts on the system performance. For example, studies from elsewhere have shown that P-solubilizing fungi are more abundant and N-fixing Bradyrhizobia more diverse and more efficient in reduced than in conventional tillage systems (Simmons and Coleman 2008; Ferreira et al. 2000). Likely for the same reasons, soybean nitrogen derived from the air (NdfA) is highest in reduced tillage plus crop residue in our experimental site (Kihara 2009).

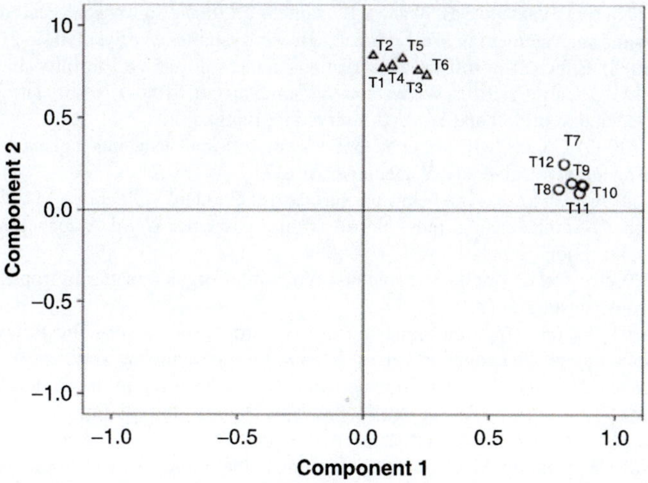

Fig. 3.9 Ordination plot of soil microbial communities in Nyabeda, western Kenya, as defined by principal components of PCR-DGGE profiles (Δ reduced tillage; o conventional tillage), T1–T12 are 12 treatments as in Table 3.2. The two principal components explained 66% of the total variance (Source: Kihara 2009)

Conclusions

Long-term application of treatments in this tropical environment has provided an opportunity for capacity building for Msc and PhD levels with most of the studies mentioned here resulting from this particular trial. A major result in most of the studies from this experiment is the increased presence and activity of soil organisms especially when reduced tillage plus crop residue is practiced. Reduced tillage with crop residue improved soil structure having the greatest aggregation at all soil depths investigated. Maize yields under the tropical Ferralsol in western Kenya are still lower for reduced tillage than conventional tillage despite the biological and physical soil improvement. Rotation of maize and soybean result in additional yield benefits relative to cropping system of maize season after season.

References

Anderson JM, Ingram JSI (1993) Tropical soil biology and fertility: a handbook of methods of analysis. CAB International, Wallingford, p 220

Ayuke FO (2010) Soil macrofauna functional groups and their effects on soil structure, as related to agricultural management practices across agroecological zones of sub-saharan Africa. Doctoral thesis. Wageningen University, Wageningen

Ayuke FO, Pulleman MM, Vanlauwe B, de Goede RGM, Six J, Csuzdi C, Brussaard L (2011) Agricultural management affects earthworm and termite diversity across humid to semi-arid tropical zones. Agric Ecosyst Environ 148:148–154

Bationo A, Kihara J, Vanlauwe B, Waswa B, Kimetu J (2007) Soil organic carbon dynamics, functions and management in West African agro-ecosystems. Agr Syst 94:12–25

Benites J (2008) Effect of no-till on conservation of the soil and soil fertility. In: Goddard T, Zoebisch MA, Gan YT, Ellis W, Watson A, Sombatpanit S (eds) No-till farming systems. World Association of Soil and Water Conservation, Bangkok

Biamah EK (2005) Coping with drought: options for soil and water management in semi-arid Kenya. Wageningen University, Wageningen, p 137

Blank D (2008) A fresh look at life below the surface. In: Goddard T, Zoebisch MA, Gan YT, Ellis W, Watson A, Sombatpanit S (eds) No-till farming systems. World Association of Soil and Water Conservation, Bangkok

Craswell ET, Lefroy RDB (2001) The role and function of organic matter in tropical soils. Nutr Cycl Agro-ecosyst 61:7–18

Curry JP (2004) Factors affecting the abundance of earthworms in soils. In: Edwards CA (ed) Earthworm ecology, 2nd edn. CRC Press, Boca Raton/London/New York, pp 91–111

Derpsch R (2008) No-till and conservation agriculture: a progress report. In: Goddard T, Zoebisch MA, Gan YT, Ellis W, Watson A, Sombatpanit S (eds) No-till farming systems. World Association of Soil and Water Conservation, Bangkok

Erenstein O (2003) Smallholder conservation farming in the tropics and sub-tropics: a guide to the development and dissemination of mulching with crop residues and cover crops. Agric Ecosyst Environ 100:17–37

Erenstein O, Sayre K, Wall P, Dixon J, Hellin J (2008) Adapting no-tillage agriculture to the conditions of smallholder maize and wheat farmers in the tropics and sub-tropics. In: Goddard T, Zoebisch MA, Gan YT, Ellis W, Watson A, Sombatpanit S (eds) No-till farming systems. World Association of Soil and Water Conservation, Bangkok

Feller C, Albrecht A, Blanchart E, Cabidoche YM, Chevallier T, Hartman C, Eschenbrenner V, Larr'e-Larrouy MC, Ndandou JF (2001) Soil organic carbon sequestration in tropical areas. General considerations and analyses of some edaphic determinants for Lesser Antilles soils. Nutr Cycl Agro-ecosyst 61:19–32

Ferreira MC, Andradeb DS, Chueirea LMO, Takemuraa SM, Hungria M (2000) Tillage method and crop rotation effects on the population sizes and diversity of bradyrhizobia nodulating soybean. Soil Biol Biochem 32:627–637

Fowler R, Rockstrom J (2001) Conservation tillage for sustainable agriculture: an agrarian revolution gathers momentum in Africa. Soil Tillage Res 61:93–107

Gachengo CN, Palm CA, Jama B, Othieno C (1999) Tithonia and senna green manures and inorganic fertilizers as phosphorus sources for maize in Western Kenya. Agrofor Syst 44:21–36

Gan YT, Goddard T (2008) Roles of annual legumes in no-till vs conventional tillage in Indiana and Illinois farm fields. In: Goddard T, Zoebisch MA, Gan YT, Ellis W, Watson A, Sombatpanit S (eds) No-till farming systems. World Association of Soil and Water Conservation, Bangkok

Gee GW, Bauder JW (1986) Particle size analysis. In: A K (ed) Methods of soil analysis Part I physical and mineralogical methods. American society of agronomy. Soil science society of American, Madison, USA

Ghuman BS, Sur HS (2001) Tillage and residue management effects on soil properties and yields of rainfed maize and wheat in a subhumid subtropical climate. Soil Tillage Res 58(1–2):1–10

Gitonga JL, Ngeru JN, Liniger HP (2008) Impacts of conservation tillage on soil water and crop production – a case study in the northwest footslopes of Mount Kenya. In: Goddard T, Zoebisch MA, Gan YT, Ellis W, Watson A, Sombatpanit S (eds) No-till farming systems. World Association of Soil and Water Conservation, Bangkok

Hurisso TT (2007) Tillage and residue management effects on soil aggregation, soil organic matter distribution and yield in a western Kenya cropping systems with and without soil fauna exclusion. Msc thesis. Wagneningen University, Wagneningen

ICRAF (1995) International centre for research in agroforestry: laboratory methods of soil and plant analysis. Nairobi, Kenya

Kihara J (2009) Conservation tillage in Kenya: the biophysical processes affecting its effectiveness. Ph.D. Thesis, University of Bonn: Bonn, Germany

Krishna K (1970) Taxonomy, phyloheny, and distribution of termites. In: Krishna K, Weesner FM (eds) Biology of termites. Academic Press, New York, London, pp 127–150
Landers JN (2008) Environmental impacts and social dimensions of zero tillage conservation agriculture in tropical Brazil. In: Goddard T, Zoebisch MA, Gan YT, Ellis W, Watson A, Sombatpanit S (eds) No-till farming systems. World Association of Soil and Water Conservation, Bangkok
Lavelle P, Brussaard L, Hendrix P (1999) Earthworm management in tropical agroecosystems. CABI, Wallingford, 300 p
Malhi SS, Lemke R, Wang ZH, Chhabra BS (2006) Tillage, nitrogen and crop residue effects on crop yield, nutrient uptake, soil quality, and greenhouse gas emissions. Soil Tillage Res 90:171–183
McCarty GW, Meisinger JJ (1997) Effects of N fertilizer treatments on biologically active N pools in soils under plow and no tillage. Biol Fertil Soils 24:406–412
Ndabamenye T (2006) Effects of soil macro-fauna, tillage and crop residue management on soil aggregation, soil organic matter dynamics and crop yields in Kenyan cropping systems. Msc thesis. Wagneningen University, Wagneningen
Okalebo JR, Gathua KW, Woomer PL (2002) Laboratory methods of soil and plant analysis: a working manual. The Tropical Soil Biology and Fertility Program, Regional Office for Science and Technology for Africa, UNESCO, Nairobi, Kenya
Rockstrom J, Kaumbutho P, Mwalley P, Temesgen M (2003) Conservation farming among small-holder farmers in E. Africa: adapting and adopting innovative land management options. In: García-Torres L, Benites J, Martínez-Vilela A, Holgado-Cabrera A (eds) Conservation agriculture. Kluwer Academic Publishers, Dordrecht/London, pp 459–470
Scopel E, Muller B, Tostado JMA, Guerra EC, Maraux F (1998) Quantifying and modelling the effects of a light crop residue on the water balance: an application to rainfed maize in Western Mexico. XVI World Congress of Soil Science. Montpellier, France, August 1998
Simmons BL, Coleman DC (2008) Microbial community response to transition from conventional to conservation tillage in cotton fields. Appl Soil Ecol 40:518–528
Taa A, Tanner D, Bennie ATP (2004) Effects of stubble management, tillage and cropping sequence on wheat production in the south-eastern highlands of Ethiopia. Soil Tillage Res 76:69–82
Terano Y (2010) Termite abundance, taxonomic richness and soil properties in conventional and conservation tillage systems in Western Kenya. Msc thesis. Wagneningen University, Wagneningen
Tursunov M (2009) Potential of conservation agriculture for irrigated cotton and winter wheat production in Khorezm, Aral Sea Basin. der Hohen Landwirtschaftlichen Fakultät. Rheinischen Friedrich-Wilhelms-Universität, Bonn, p 106
Wander MM, Yang X (2000) Influence of tillage on the dynamics of losse and occlude particulate and humified organic matter fractions. Soil Biol Biochem 32:1151–1160

Chapter 4
Strategies for Maintenance and Improvement of Soil Productivity Under Continuous Maize and Beans Cropping System in the Sub-humid Highlands of Kenya: Case Study of the Long-Term Trial at Kabete

C.N. Kibunja, F.B. Mwaura, D.N. Mugendi, P.T. Gicheru, J.W. Wamuongo, and A. Bationo

Abstract Inappropriate soil management practices are cited as the main causes of soil fertility decline of cultivated lands in sub-Saharan Africa (SSA). Small-scale farmers are faced with daunting challenges and limited opportunities for maintaining the productivity of their land due to low accessibility and affordability of agricultural inputs. In Kenya, a series of fertilizer trials conducted throughout the country in the 1990s, showed that nitrogen (N) and phosphorus (P) were deficient in 57% and 26% of the sites covered. The effect of continued cultivation with application of mineral and organic fertilizers on soil quality and crop yields was studied in a long-term field experiment at Kabete, in the highlands of Kenya, which was started in 1976. The area is sub-humid with an average bimodal rainfall of 980 mm and two cropping seasons per year. The treatments included: control (no-input); fertilizer nitrogen (N) and phosphorus (P) at 60 kg N/ha and 26 kg P/ha, and 120 kg N/ha and 52 kg P/ha; farmyard

C.N. Kibunja (✉) • P.T. Gicheru
National Agricultural Research Laboratories, Kenya Agricultural Research Institute, P.O. Box 57811–00200, Nairobi, Kenya
e-mail: catherine.kibunja@yahoo.com; cdnarl@iconnect.co.ke

F.B. Mwaura
School of Biological Sciences, University of Nairobi, P.O. Box 30197, Nairobi, Kenya
e-mail: fbmwaura@uonbi.ac.ke

D.N. Mugendi
School of Environmental Sciences, Kenyatta University, P. O. Box 43844, Nairobi, Kenya
e-mail: dmugendi@yahoo.com

J.W. Wamuongo
Natural Resources Management, Kenya Agricultural Research Institute, P.O. Box 57811–00200, Nairobi, Kenya
e-mail: JWWamuongo@kari.org

A. Bationo
Soil Health, Alliance for a Green Revolution in Africa (AGRA), Accra, Ghana
e-mail: ABationo@agra-alliance.org

A. Bationo et al. (eds.), *Lessons learned from Long-term Soil Fertility Management Experiments in Africa*, DOI 10.1007/978-94-007-2938-4_4,

manure (FYM) at five (5) and ten (10) t/ha and farmyard manure (5 t/ha) combined with 60 kg N and 26 kg P/ha. For each treatment, crop residues were retained (+) or removed (–) in half of the plots. Maize and beans were planted during the long and short rain seasons, respectively. This paper presents a review of experimental results derived from the long-term experimental site at Kabete, Kenya over a period of 30 years. Results indicate that application of all inputs significantly ($p = 0.05$) improved maize yields over the control but there was no significant differences among the treatments. Cumulative use of FYM gave better yields than NP fertilizers but the combined use of FYM and inorganic fertilizers was the most economically and promising strategy for sustained soil crop yields A general decline in soil bio-physical properties was observed over time. Soil acidification, decline in soil organic carbon and nitrogen has been noted over time. Use of inorganic fertilizers without addition of FYM or crop residues led to loss of soil biodiversity. None of the strategies used have maintained carbon stocks at the initial level. Integrated use of farmyard manure and chemical fertilizers, combined with liming to reduce soil acidification and use of deep rooted rotational crops should be recommended to sustain soil productivity under continuous cultivation.

Keywords Long-term • Maize-bean rotation • NP fertilizers • Manure • Crop residues

Introduction

Continuous cultivation of land due to increasing population pressure has led to serious soil fertility decline in sub-Saharan Africa (SSA) as manifested by declining crop yields, decreasing vegetation cover and increasing soil erosion. As a result, farm productivity and agricultural incomes are falling; migration to urban centres is on the increase while both household and countrywide food security is declining (FAO 1995). In general the SSA region faces two interrelated problems: declining per capita production and progressive deterioration of the environment (Scholes et al. 1994). The sub-Saharan Africa is still the only region in the world where per capita food production has stagnated over the last 40 years leading to acute food insecurity. This is due to a culmination of poor farming practices which have led to serious soil fertility decline, low crop yields, decreased ground cover and progressive environmental deterioration (FAO 1995). Inherent low soil fertility and low use of organic/mineral fertilizers are among the main biophysical constraints to increasing agricultural productivity in sub-Saharan Africa. Replenishment of soil fertility is cited as the most important activity associated with enhanced crop production (Sanchez et al. 1996).

In large parts of Kenya, food crop yields are very poor due to declining soil fertility as a result of continuous cropping with little or no replenishment of the nutrients removed through either crop harvests or other losses such as leaching and soil erosion (Kibunja 2007). Small-scale farmers are faced with daunting challenges and limited opportunities for maintaining the productivity of their land. Shortage of capital and poor access to credit facilities often hampers the use of inorganic fertilizers (Minae and Akyeampong 1988). The current economic policy reforms have led to high fertilizer prices compared to crop harvests leading to low inputs: output returns and stagnation in agricultural productivity (Heerink 2005). Continuous land sub-division reduces the opportunity for fertility restoration through fallowing. This is especially true of the central highlands, which form an important agricultural zone of Kenya (Siderius and Muchena 1977). The area enjoys a relatively good and reliable rainfall with fertile soils, and supports a rapidly increasing population. Continuous cropping, with little or inadequate inputs, causes heavy nutrient losses through the harvested food and cash crops which are exported out of the system (Stoorvogel et al. 1993). This in combination with soil erosion and leaching has lead to widespread "mining" of soil nutrients and depletion of soil fertility causing a serious threat to food security in the country.

Nitrogen, next to water, is the most limiting element to agricultural productivity (Sanchez 1976). It is also the most widespread limiting nutrient to agriculture in Africa followed by phosphorus (Bekunda et al. 1997; Okalebo 2003). In Kenya, a series of fertilizer trials conducted throughout the country, showed that nitrogen (N) and phosphorus (P) were deficient in 57% and 26% of the sites, respectively (Kenya Agricultural Research Institute (KARI) 1994). The main sources of plant-available nitrogen are through mineralization of soil organic matter, biological nitrogen fixation, chemical fertilizers and organic inputs such as plant residues, composts and manure. Application of animal manures is a popular method of maintaining soil productivity in eastern and southern Africa where up to 83% of farmers apply animal manure on their cropland (Simpson et al. 1996). According to a survey conducted among 168 farmers in four districts of Kenya, though both fertilizers and farmyard manure are used, the quantities often used are below the recommended rates (Kapkiyai et al. 1998) due to a scarcity of organic resources and inability to purchase chemical fertilizers. The quantities of available manure are also low due to small herd sizes kept by these farmers. Crop residues are not normally returned to the field but fed to confined animals, which have been reported to produce only about 1,500 kg manure cow^{-1} $year^{-1}$ (Strobel 1987). Probert et al. (1992) further reported that in the semi-arid parts of Kenya, the type of manure often used is of very low quality [mean of 0.42% N; 0.17% P; 0.9% Potassium (K)] and therefore not enough to replace plant nutrients lost through harvested crops. Qureshi (1991) reported that one maize crop (both grain and stover) removed a total of 91 kg N, 16 kg P and 139 kg K from the soil. Most K removed was in the stover, most P in the grain while the N removed was found equally divided between the grain and stover. Further computation of the above data, showed that about 20–40 kg N, 2–4 kg P and 15–30 kg K was extracted from the soil by each 1 t ha^{-1} crop of maize dry matter harvested (Simpson et al. 1996). These figures are in agreement with the negative balances derived for similar agro-ecological zones by Stoorvogel et al. (1993). However, manure still remains a valuable source of bases (Probert et al. 1992) and also improves the water-holding capacity of soils in the short term. Other options used by farmers in this region to alleviate soil fertility problems include, the use of biological materials such as legume cover crops, legume intercrops and rotations, use of plant litter, crop residues, dead roots, compost, household and industrial wastes although not strictly as an alternative to mineral fertilizers.

Status of Agriculture in Kenya

Agriculture is the backbone of Kenya's economy contributing directly 26% of gross domestic product (GDP) and indirectly 60% of the export earnings and is thus crucial to the country's overall economic and social development. The feeding of Kenya's growing population will require intensification of agricultural production in the medium and high rainfall areas under rain fed conditions and expansion of irrigation in the lower rainfall areas. Sustenance of agricultural production will therefore call for successful management of soil and water resources to satisfy home needs, while maintaining or enhancing the quality of the environment and conserving these resources. Kenya has experienced a persistent decline in agricultural growth since the 1980s leading to low crop productivity, chronic food shortages and rising poverty levels (Scholes et al. 1994; FAO 1995). This has been attributed to low inherent soil fertility combined with inappropriate soil management practices (Sanchez et al. 1996). Continuous cropping combined with inadequate soil nutrient replenishment has led to serious soil fertility decline as manifested by declining crop yields, decreased vegetation cover, increased soil erosion and siltation of rivers and dams, decreased water quality, increased incidences of pests and diseases, high incidences of noxious weeds such as *Striga hermonthica*, potential increases in greenhouse gases with consequent effects on climate, and increased prevalence of catastrophic events such as landslides and floods (Onduru et al. 2006).

Soil fertility related issues are a major concern to Kenya as attested by numerous policy documents and research projects, e.g., the Strategy to Revitalize Agriculture policy document of the Kenya Governemen 2004 states that "*low and declining fertility of the land*" is one of the factors that continue to limit the growth of agriculture in the country. This has led to declining crop yields, poor soil physical properties such as decreased soil aggregation, bulk density and water infiltration, loss of vegetative cover and biological diversity (Onduru et al. 2006). Tackling soil fertility issues thus requires a long-term perspective and holistic approach of the kind embodied in the concept of Integrated Soil Fertility Management (ISFM). Thus ISFM is the adoption of a holistic approach to research on soil fertility that embraces the full range of driving forces and consequences (biological, physical, chemical, social, economic and political) of soil degradation.

Long Term Experiment at Kabete, Kenya

Early soil fertility work in East Africa started in the 1930s and addressed the issue of soil fertility restoration through the combined use of vegetative fallows and animal manures. This was in response to the slowly emerging scenario of decreased shifting cultivation accompanied by more settlement and population increases. Traditional farming systems in SSA were supported by shifting slash-and-burn cultivation, a low input sustainable agricultural farming system that allowed the land to rest for several years to restore fertility (Stangel 1995). In these systems, slash and burn provided adequate nutrients from bush and tree ashes while the long fallow periods (15–20 years) allowed the soil to recuperate. However, as population pressure increased and raised the demand for food, feed and fibre, the interval between the fallows were substantially shortened. This resulted in overall decline in land productivity as a result of reduced soil fertility and particularly soil organic matter content (Stangel 1995).

In East Africa, several studies were carried out to find ways of replenishing nutrients and boosting land productivity due to the limitations of the traditional farming practices accompanied by decreasing cropland (Simpson et al. 1996). The studies highlighted the importance of organic sources of nutrients in enhancing nutrient availability to the growing crop. A lot of interest was also created in studies on the dynamics of soil N mineralization cycles in the 1950s (Birch 1958). In the 1960s, studies were concerned with the causes of soil fertility decline under continuous cropping such as leaching of nutrients into the subsoil beyond the reach of the annual crops grown (Jones 1967). In Uganda, Jones (1972) reported that yields of beans in the control treatment declined steadily over a period of 4 years whereas yields of beans receiving fertilizers were constant over the same period. These results further showed that it was possible to reduce the fallow period to 1 year and also replace the natural soil fertility lost during an arable phase by the routine use of inorganic fertilizers (Jones 1972). This arrangement appeared to offer the possibility of continuous cropping without a resting period. The research on the intensive use of fertilizers nonetheless failed to have much impact on the small-scale farming community in the medium potential areas largely due to socio-economic constraints. However, it led to increased use of fertilizers in large-scale farming in the East African region and the adoption of hybrid maize varieties in the high potential areas, e.g., western Kenya. As a result of these developments, several long-term soil fertility trials were set up in Africa to investigate the effects of different nutrient management practices on soil fertility and crop yields.

One of the major advantages of long-term fertility trials is that they enable the monitoring of soil nutrient changes over a long period of time which may not be captured by short-term experiments. According to a review by Swift et al. (1994), 21 long-term experiments were documented in Africa between 1937 and 1990. Three of these experiments were in existence for more than 20 years, e.g., the trial in Serere, Uganda (1937–1964), the Gezira trial in Sudan (1941–1966), and the Saria trial in Burkina Faso, which was set up in 1960. Of these, only one is still operative, that is, the Saria trial in Burkina Faso. Other trials of shorter duration exists, for example, the one in Kabete, Kenya, which was started in 1976 (Qureshi 1979) and is still in existence. The experience of these long-term experiments in Africa has shown the role of nutrients derived from organic inputs in maintaining yield levels. The yield decline is almost always related to organic matter decline (Swift et al. 1994; Bationo et al. 2007; this publication).

This paper will present a review of experimental results derived from the long-term experimental site at Kabete, Kenya over a period of 30 years.

Methodology

Site Description, Experimental Design and Management

The Kabete long-term soil fertility improvement and maintenance trial is located at National Agricultural Research Laboratories (NARL), of the Kenya Agricultural Research Institute (KARI) at Kabete, Kenya. It was established in 1976 to provide sustainable methods of sustaining soil fertility and land productivity. The primary objective of the experiment was to find appropriate methods of maintaining and improving the productivity of soil through repeated use of inorganic nitrogen (N) and phosphorus (P) fertilizers, farmyard manure (FYM) and crop residues (R) under continuous maize-bean rotation following the practices of the small-scale farmers (Hinga et al. 1980; Qureshi 1979). The site lies at longitude 36°46′E, and latitude 01°15′S and is located about 8 km Northwest of the city of Nairobi, at 36° 41′E and 01° 15′S and at an altitude of 1,787 m above sea level in the upper semi-humid agro-ecological zone UM3 (Jaetzold et al. 2006). The site has two rainy seasons each year; the first (mid-March to May) and the second (mid-October to December) receiving 409 and 220 mm of rainfall, respectively (Jaetzold et al. 2006). The mean annual temperature ranges from 18°C to 21°C. The soil has been classified as Humic Nitisol (UNESCO 1974) or as Ustic Tropohumult (USDA 1975) and is known locally as the Kikuyu Red Clay which is considered to be of moderate fertility (Siderius and Muchena 1977). The soil is clayey in nature and the site had an initial natural vegetation of grassland and closed savannah. The soil type represents an important agricultural zone in Kenya i.e., the Central Kenya and Kisii regions covering a total of about 55,000 ha of land in Kenya and about 228,000 sq km within the E. African highlands (Fig. 4.1).

The experimental procedures consists of 18 treatments laid out in a randomized complete block design with four replications in various combinations of inorganic fertilizers (N, P), farmyard manure and crop residues in a 2×3×3 factorial arrangement as shown in Table 4.1. Treatments included three different levels of inputs while crop residues from the preceding season's crop were either incorporated in the soil or removed. The cropping system consists of a maize crop during the long-rains (LR) season rotated with beans during the short rains (SR). The site has been maintained for the last 35 years with little variation in treatments.

The plot sizes measured 7.0 m×4.5 m covering a total area of 31.5 m^2 and were distributed in four replicate blocks. Chemical NP fertilizers and manure were applied once a year during the LR season. Phosphate fertilizer was incorporated into shallow furrows at the rate of 0, 26 and 52 kg P/ha as triple super-phosphate (TSP) at planting. Farmyard manure (FYM) was applied at three rates (0, 5 and 10 t/ha) in shallow furrows at least 2 weeks before planting. Nitrogen fertilizer was applied at three rates (0, 60 and 120 kg N/ha) as calcium ammonium nitrate (CAN) 7 weeks after planting maize in one application; however, since the year 2008, the CAN was applied in two spilt applications, at 4 and 6 weeks after planting. Crop residues were chopped and incorporated into the soil at the time of land preparation. Land

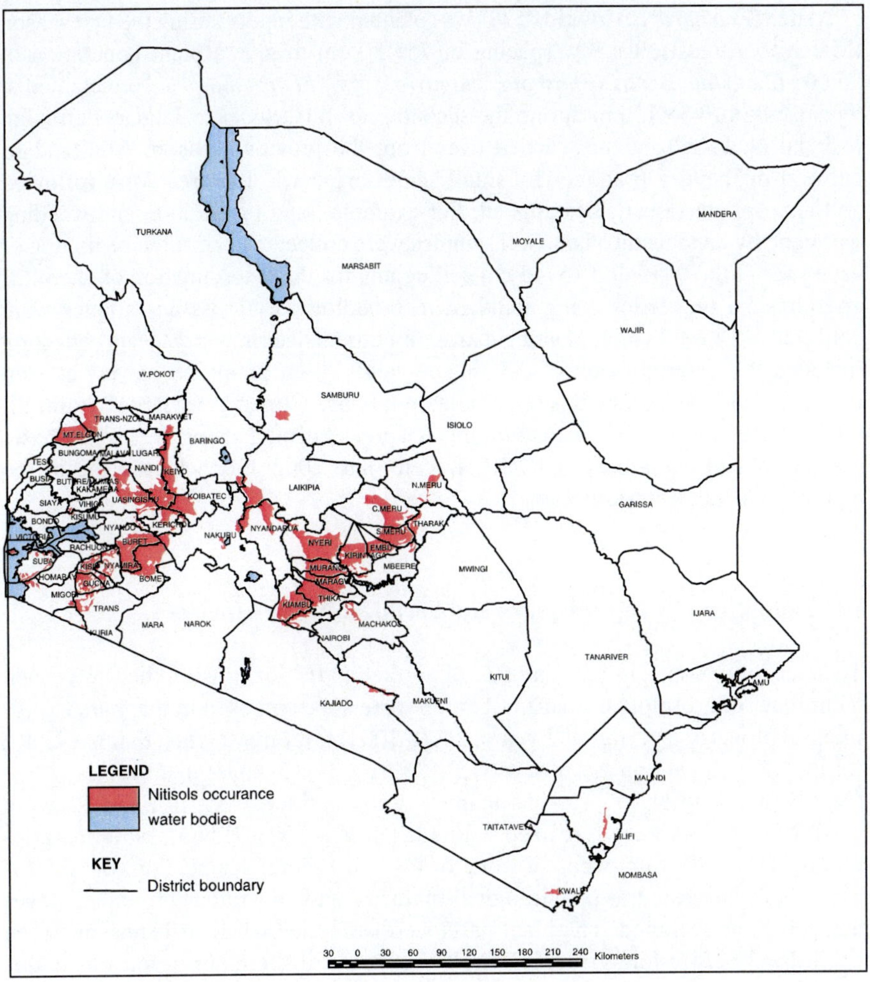

Fig. 4.1 Map showing the distribution of Nitisols in Kenya (Source: Kenya Soil Survey 2001)

Table 4.1 Experimental treatments applied in the long-term experiment at NARL, Kabete, Kenya

Agricultural inputs	Level 1	Level 2	Level 3
N, P fertilizers (kg/ha)	0,0	60,26	120,52
Farmyard manure (FYM t/ha)	0	5	10
Crop residues (R)	+	–	N/A

Legend:
N Nitrogen, *P* Phosphorus, *R* Crop residues

preparation, planting and weeding was done manually using a hand-hoe. Beans were grown on residual fertilizers utilizing residual N, P and manure carried over from the previous season.

Maize (*Zea mays* L.) hybrid "*512*" was planted with inputs during the first season (March to August) at a row spacing of 75×50 cm to give a plant population of 55,000 plants/ha. Beans (*Phaseolus vulgaris* L. cv *"Mwezi moja"*) were planted at the spacing of 45×15 cm during the second season (October to January) utilizing residual N, P and manure carried over from the previous season. All standard cultural husbandry practices by small scale farmers in the area were followed and appropriate insecticides applied. For example, land preparation and weeding was done by hand using a hoe. Soil samples were collected once at the beginning of each year at the 0–15 and 15–45 cm soil depths for the determination of chemical and physical properties using standard methodology. Plant tissue samples were collected at 12 weeks after planting maize for nutrient determination. Data collection included the determination of dry matter yields (both grain and stover) at crop maturity, soil and plant tissue samples analysis. Climatic variables (rainfall, temperature, evaporation and humidity) were measured using the some basic meteorological equipment at KARI, Kabete upto 2002 after which some of the equipment became unserviceable.

Determination of Crop N Uptake and Fertilizer N Use Efficiency

To determine crop N uptake and fate of applied N fertilizer using the ^{15}N isotope technique, micro-plots measuring 1×2 m were superimposed in the main experimental plots for a period of 2 years, 2000–2001. Four treatments, that is, $N_{60}\ P_{26}$ (N_1P_1); $N_{60}\ P_{26}$+Crop residues (N_1P_1R); $N_{60}\ P_{26}$+5 tons manure ($N_{60}\ P_{26}$) ($N_1P_1FYM_5$) and $N_{60}\ P_{26}$+5 tons manure+Crop residues ($N_{60}\ P_{26}FYM_5R$) were used. Micro-plots were fertilized with a 0.1 M Ca $^{15}NH_4$ ($^{15}NO_3)_3$ solution containing 10% ^{15}N atom excess at time of top – dressing. Maize from the micro-plots were harvested at physiological maturity and separated into cobs, stover and roots and weighed. The plant parts were shredded when still fresh and then oven-dried before fine-grinding. The ^{15}N content of the plant tissue was determined using an emission spectrometer (NO-I PC, GmHB, Germany) as outlined in IAEA (2001). Fertilizer N utilization or recovery by the crop is the fraction of the fertilizer N taken up by the plant in relation to the rate of fertilizer N applied and was finally calculated as:

$$\%\text{N recovery} = \frac{\text{Fertilizer N yield}}{\text{Rate of N application}} \times 100 \qquad (4.1)$$

To study the dynamics of mineral N in the soil profile, soil samples were collected at 0–20; 20–50; 50–100; 100–200 and 200–300 cm three times per season (at planting; at flowering and after harvesting) using a 5 cm Edelman auger. Mineral nitrogen (NH_4-N+NO_3-N) was determined by the KCl extraction method while total nitrogen was determined by the modified Berthelot reaction using an Automated 2 – Channel Segmented Flow Analyzer (Model SA 4000) by Skalar, Netherlands (Houba et al. 1988; Novozamsky et al. 1983).

Determination of Changes in Soil Bio-Physical Properties Over Time

In order to assess the changes in soil quality after continuous cropping for 25 years, soil bio-physical parameters were determined and compared with the values of the original samples collected in 1976. Soil sampling was done at the beginning of the maize planting season at 0–15 cm and 15–45 cm depths just before land preparation in the years 2000 and 2001. Soil nitrogen, organic carbon content, soil acidity and bulk density was determined using standard methodology (Mehlich et al. 1962; Hinga et al. 1980).

Data Analyses

All the crops were harvested at maturity and grain yields were expressed at 12.5% moisture content. The data was subjected to analysis of variance (ANOVA) using GENSTAT version 5 (Genstat 5 Committee 1993). The least significance difference (LSD) was used to compare treatment means and treatments declared different at $P = 0.05$ level of significance, unless otherwise stated.

Research Results

Crop Yields

Interesting results have been generated over the years, which can be transferred into technologies for immediate transfer to stakeholders. For example, during the first 5 years, the use of chemical fertilizers alone increased maize yields by about 50% (4–6 $t\,ha^{-1}$) and out-yielded all other treatments. However, this trend changed soon after the sixth year when the treatments with combined chemical fertilizers and farmyard manure out-yielded all the other treatments. The significance of long-term experimentation was demonstrated here because had the trial been stopped after 5 years, this insight would have been lost and wrong recommendations would have been given out. To-date the treatments with integrated inorganic and organic inputs (NP and manure) consistently gave better yields than treatments with inorganic inputs (NP) alone as shown in Fig. 4.2. Maize grain yields in the unamended control declined from 4.0 to 2.0 $t\,ha^{-1}$ within the first 10 years and currently stands at about 1.0 $t\,ha^{-1}$.

Cumulative application of farmyard manure (FYM) gave better yields than chemical fertilizers (NP) even at the higher level of 120 N and 52 P kg/ha (Fig. 4.3). All inputs applied significantly ($p = 0.05$) improved yields over the control but there was no significant differences among the treatments. Integrated use of manure (>5–10 t/ha) and inorganic fertilizers (>60 N + 30 P kg/ha) was the most economically and promising strategy for SOM build up and sustained soil crop yields in

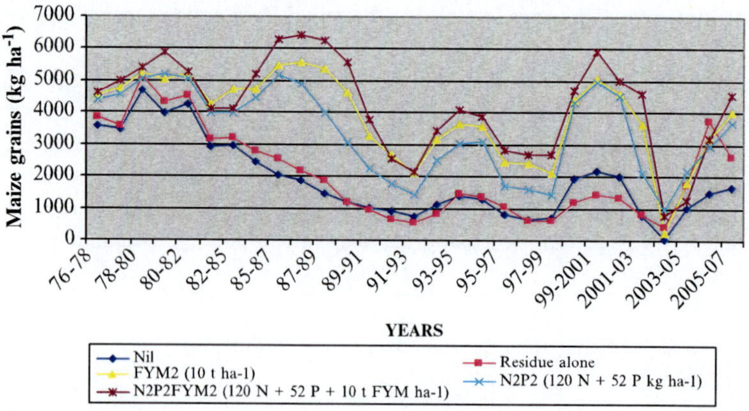

Fig. 4.2 Average yields for maize grains at the Kabete long-term experiment from 1976 to 2007 (Source: Kibunja 2007)

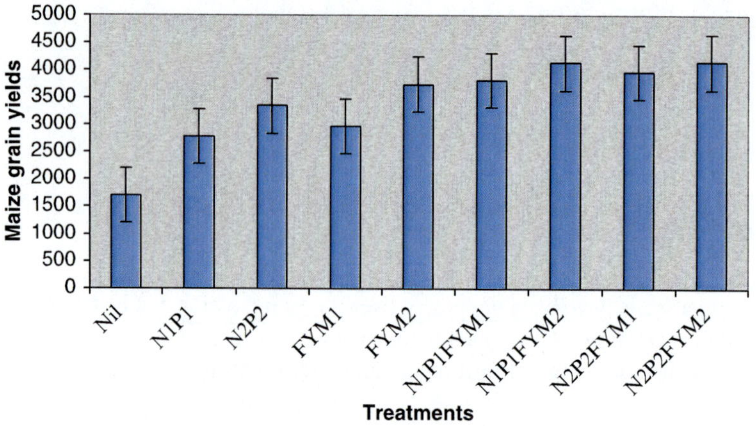

Fig. 4.3 Average maize yields using different management options over a period of 30 years at Kabete, Kenya. Legend: Nil – control; N1P1 – 60 kg N ha^{-1} + 26 kg P ha^{-1}; N2P2 – 120 kg N ha^{-1} + 52 kg P ha^{-1}; FYM1 – farmyard manure (FYM) at 5 tha^{-1}; FYM2 at 10 tha^{-1}; N1P1 FYM1 – 60 kg N ha^{-1} + 26 kg P ha^{-1} + 5 tha^{-1} FYM; N1P1 FYM2 – 60 kg N ha^{-1} + 26 kg P ha^{-1} + 10 tha^{-1} FYM; N2P2 FYM2 – 120 kg N ha^{-1} + 52 kg P ha^{-1} + 10 tha^{-1} FYM

the cropping system. Additionally, evidence of profitability is also exhibited by the results of the 35 year old, long term trial which has demonstrated yield responses to inorganic, organic fertilizers and their integration. These results also infer that in terms of crop yields, it may be more useful to feed stover to cattle and return manure to the crop field.

Detailed analysis of maize yields over the last 25 years using mixed model equation was also carried out to show the most sustainable and cost-effective management options for the small-scale farming community. All management strategies

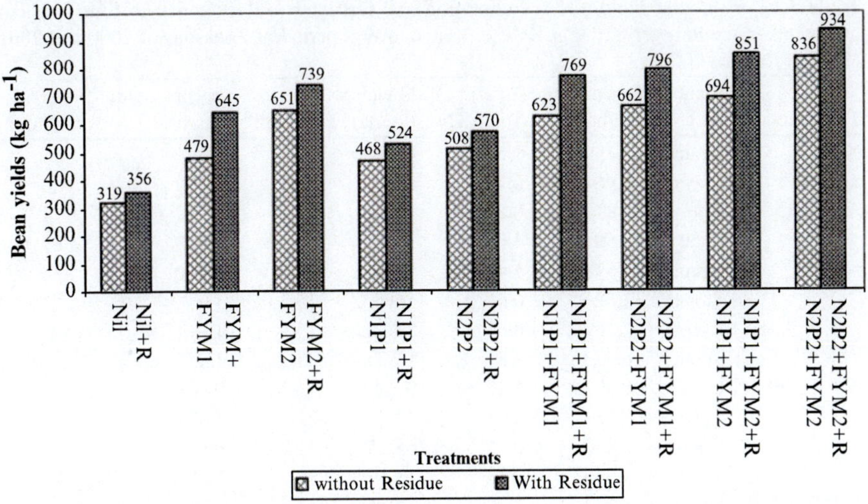

Fig. 4.4 Effect of maize stover restitution on average bean yields under continuous cropping from 1976 to 1998 (Kibunja et al. 2003). Legend: Nil – control; N1P1 – 60 kg N ha^{-1}+26 kg P ha^{-1}; N2P2 – 120 kg N ha^{-1}+52 kg P ha^{-1}; FYM1 – farmyard manure at 5 t ha^{-1}; FYM2 at 10 t ha^{-1}; N1P1 FYM1 – 60 kg N ha^{-1}+26 kg P ha^{-1}+5 t ha^{-1} FYM; N1P1 FYM2 – 60 kg N ha^{-1}+26 kg P ha^{-1}+10 t ha^{-1} FYM; N2P2 FYM2 – 120 kg N ha^{-1}+52 kg P ha^{-1}+10 t ha^{-1} FYM

that included chemical fertilizers posted negative returns perhaps due to the high cost of the fertilizers compared to the cost of the produce. However, the application of 10 t ha^{-1} FYM gave higher economic returns compared to other treatments (Thuranira 2007).

Further results indicate that continuous cropping of beans with residual or no inputs gave consistently low and declining yields (Qureshi 1991), which were far below the optimum yields of 1,000–2,000 kg ha^{-1} reported for this region (Keya 1977; KARI 1994; Chui 1994). The yields ranged from 600 to 950 kg ha^{-1} at the beginning of the field experiment (Qureshi 1979), but fluctuated from 300 to 1,000 kg ha^{-1} during the first 23 years depending on the management option and amount of rainfall received (Kibunja and Gikonyo 2000; Kibunja et al. 2003). Beans were grown on biologically fixed N with no additional inputs except maize stovers in some selected plots. Beans benefitted from maize stover application as indicated by average bean yields calculated over a period of 23 years (Fig. 4.4). Currently, the average bean yields have declined from 950 kg ha^{-1} to about 300–500 kg ha^{-1} probably due to increasing soil acidity, declining soil fertility and high incidence and diseases.

Fertilizer N Recovery

Fertilizer N recovery for four seasons (two maize crops and one bean crops) in the period 2000–2002 is shown in Table 4.2. The total dry matter (grains and stovers)

Table 4.2 Dry matter yields, N accumulation and fertilizer use efficiency in four selected treatments in the long-term trial at Kabete, Kenya, over a period of 2 seasons of 2000 and 2001 (Source: Kibunja 2007)

Treatments	Plant part	Dry matter (kg/ha)	Total %N	N yield (kg/ha)	% Ndff†	Fert N yield (kg/ha)	% Fert N utilization
N_1P_1	Grain	2,750	1.57	4316.7	14.5	626	
	Stover	8,964	0.71	6364.2	18.5	1,177	
	Roots	878	0.80	702.4	17.9	125.7	
	Total	12,592	3.08	11,383	50.9	1,929	32.15
N_1P_1R	Grain	3,538	1.49	5271.6	12.5	659.0	
	Stover	11,526	0.66	7607.0	15.3	1164.0	
	Roots	1,269	0.65	824.8	15.9	131.1	
	Total	16,333	2.8	13,703	43.7	1,954	32.5
$N_1P_1FYM_1$	Grain	6,003	1.40	8404.2	11.2	941.2	
	Stover	6,723	0.63	4235.5	12.2	516.7	
	Roots	850	0.65	552.5	11.8	65.2	
	Total	13,576	2.68	13,192	35.2	1,523	25.4
$N_1P_1FYM_1R$	Grain	5,692	1.38	7,854	11.6	911.0	
	Stover	7,249	0.65	4711.9	14.2	669.1	
	Roots	654	0.72	470.1	12.1	56.9	
	Total	13,595	2.75	13,036	37.9	1,637	27.2

Legend:
Nil – control; N_1P_1 – 60 kg N ha^{-1}+26 kg P ha^{-1}; N_2P_2 – 120 kg N ha^{-1}+52 kg P ha^{-1}; FYM_1 – farmyard manure (FYM) at 5 tha^{-1}; FYM_2 at 10 tha^{-1}; N_1P_1 FYM_1 – 60 kg N ha^{-1}+26 kg P ha^{-1}+5 tha^{-1} FYM; N_1P_1 FYM_2 – 60 kg N ha^{-1}+26 kg P ha^{-1}+10 tha^{-1} FYM; N_2P_2 FYM_2 – 120 kg N ha^{-1}+52 kg P ha^{-1}+10 tha^{-1} FYM; † – % Ndff denotes % Nitrogen derived from fertilizer

and the total N yield were higher in the treatments with combined inputs compared to treatments with fertilizer alone although not significantly different. However, the total amount of N derived from fertilizer (fertilizer N yield) was higher in the treatments with chemical fertilizers alone (32.7 kg ha^{-1}) than in treatments with combined inorganic and organic fertilizer inputs (23.7–30.2 kg ha^{-1}).

While the integrated treatment with both organic and inorganic fertilizers gave the highest grain yields, plant N content in the grain, stover and roots was highest in the plots with chemical fertilizers alone (Table 4.2). The ratio of grain: stover yield was 1:1 in the treatments where both organic and inorganic inputs had been applied but 1:3 in the treatments with inorganic fertilizers alone. Similarly, the amount of N derived from fertilizer (% ndff) was higher in the fertilizer alone treatments (12.5–14.5%) than in the combined inputs (11.2–11.6%). This is probably because in the combined input treatments, the plant had access to other sources of N. However, the fertilizer N utilization was low at 25–32% for all treatments although it was slightly better where chemical fertilizers alone were applied but there was no significant difference between the treatments. The recovery rates were comparable to those reported in literature (Mughogho et al. 1986; Haggar et al. 1993; Harris et al. 1994) which ranged from 20% to 37%. According to a review by Mughogho et al. (1986), the ^{15}N recoveries for various cereal-based cropping systems in sub-Saharan countries ranged from 27% to 45%. The studies further showed that

Table 4.3 Effect of different land management options on soil organic carbon and nitrogen content in the top-soil (0–20 cm) at Kabete, Kenya, after continuous cropping for 25 years (Kibunja 2007)

	Treatments[b]					
		2002				
Properties	1976[a]	NIL	N_1P_1	FYM_1	$N_1P_1FYM_1$	$N_1P_1FYM_1+R$
pH (1:2 in H_2O)	5.50[b]	4.20[a]	4.27[a]	4.45[a]	4.35[a]	4.40[a]
SOC (%)	2.00[c]	1.26[b]	0.93[a]	1.26[b]	1.44[b]	1.43[b]
Total N	0.16[a]	0.12[a]	0.14[a]	0.15[a]	0.15[a]	0.16[a]
Soil texture (%)	← 67% clay: 22% silt: 11% sand →					
Textural Class	← Clay →					
Bulk density (g cm^{-2})	1.04[a]	1.08[a]	1.08[a]	1.06[a]	1.04[a]	1.04

a†1976 Source of data: Qureshi (1979)

[b]Treatment key

Nil – control; N_1P_1 – chemical fertilizer at 60 kg N ha^{-1} + 26 kg P ha^{-1}; FYM_1 – farmyard manure at 5 tha^{-1}; N_1P_1+FYM_1 – 60 kg N ha^{-1} + 26 kg P ha^{-1} + 5 tha^{-1} FYM; N_1P_1+FYM_1+R – 60 kg N ha^{-1} + 26 kg P ha^{-1} + 5 tha^{-1} FYM + stover returned. *Means followed by the same letters are not significantly different at the p=0.05 level*

the application of N fertilizers to tropical soils enhanced mineralization and crop exploitation of native soil N reserves leading to low fertilizer N recoveries (Mughogho et al. 1986). In general, the recovery of fertilizer N in the biomass was low and rarely greater than 50–60% of that applied, even in the developed countries where very high fertilizer rates were applied (Mughogho et al. 1986; Haggar et al. 1993).

Changes in Soil Properties

A general decline in soil bio-physical properties was observed after continuous cropping for 25 years as shown in Table 4.3.

Soil pH

A highly significant increase of between 0.13 and 0.43 pH units was recorded in the topsoil (0–15 cm) within the first 10 years with an exception of plots receiving fertilizer N inputs (N_2P_2) that resulted in acidification. Acidification could have been due to the substantial removal of bases through grain and stover off-take coupled with leaching of cations (Hartemink et al. 1996). Such removal of cations may have contributed to rapid acidification of plots receiving mineral fertilisation only. Fertilizer application significantly ($p<0.01$) reduced soil pH while FYM and maize stover significantly ($P<0.001$) increased soil pH compared to the control. The various interactions of fertilizer, manure and stover were not significant compared to their corresponding treatments alone. Fertilizer caused a decrease in pH possibly due to a reduction in the buffering capacity of the soil. Organic materials on the other hand, had an ameliorative effect on the soil in that they increased the soil buffering capacity. Effective CEC increased significantly through addition of both organic and inorganic fertilizers in the order FYM > NP fertilizers > Crop Residues. Hence, replenishment of such cations through organic inputs with or without mineral fertilizers, apparently helped minimise or eliminate the process of acidification. Currently, the soil pH has declined significantly ($p=0.05$) by 1.0–1.3 units (Table 4.3).

From correlation analysis, both soil pH and ECEC showed a strong relationship with SOM. At the 0.1% level of significance, the correlation coefficients for pH and ECEC on total organic carbon were 0.67 and 0.74, respectively while for total nitrogen the correlations were 0.74 and 0.84, respectively (Kapkiyai et al. 1998). Soils receiving fertilizer alone showed significant ($p<0.01$) reduction in pH while manure and stover treatments showed significant ($p<0.001$) increases in soil pH. Soil organic matter is often an important source of negative charge in these soils and as such promotes retention of available forms of base cations. It is clear therefore that destruction of SOM will reduce CEC levels and have a negative impact on base cation retention.

Soil Organic Carbon

Results from this trial indicate that over the past 30 years, soil organic carbon (SOC) content has significantly declined in all plots (Table 4.3) suggesting an imbalance between SOC inputs and losses from soil, with likely effects on soil fertility and crop productivity (Swift et al. 1994; Woomer and Muchena 1996). After 18 years of different soil fertility management strategies, Kapkiyai et al. (1998) found significant differences in SOC stocks attributed to the different land management strategies. A decrease in SOC ranging from 30% in the integrated fertilizer, manure and stover managed plots to 42% in the no-input control plots was reported. Manure application had the greatest effect on soil chemical properties causing significant increases (50%) on SOC, particulate organic matter (POM), soil microbial biomass, SOM and crop yields (Kapkiyai et al. 1998). The least soil C was lost from the treatment receiving fertilizer, manure and stover. Compared to the control, various inputs increased the C content of the soil by 3–27%, with fertilizer treatment giving the lowest increment and a combination of manure and crop residues showing the highest increment. Only FYM increased total SOC significantly ($p<0.001$) indicating clearly the positive effect of FYM on the organic matter level of the soil. None of the treatments had a significant affect on non-labile C but the labile C may be a more sensitive indicator of the C dynamics of the system (Kapkiyai et al. 1998).

Further studies in the same site showed that the mean soil organic carbon concentration in the soil plough layer of 21 g C/kg soil declined by 24% in control plots and 19% in plots that received inorganic fertilizer (N1P1 and N2P2) after only 22 years of continuous cropping (Ayaga 2000). Generally the use of 10 tons of FYM/ha alone or in combination with inorganic fertilizers and crop residues apparently sustained soil organic carbon at significantly higher levels ($p<0.05$) compared to the control. However the continuous use of inorganic fertilizer alone did not arrest the decline in SOC compared to the control plots. According to Ayaga 2000, the retention of crop residues maintained the amount of soil organic C at slightly higher levels than the control, although the difference between the two mean C concentrations were not statistically significant ($p<0.05$).

All these results indicate that continuous cultivation had an adverse effect on soil organic carbon stocks which declined by 63% from 2% in the no-input control plots after 30 years of continuous cropping with no agricultural inputs (Kibunja 2007). Soil organic carbon (% SOC) content showed a general decline of about 28–37% regardless of the management option over the last 30 years (Table 4.3). Soil organic carbon now ranges from 0.93% in the treatments with chemical fertilizers to 1.44% in treatments with combined chemical fertilizers and farmyard manure. The decline was highest in the plots where chemical fertilizers alone were used (53%), followed by the no-input control plots (37%) and was lowest in the plots where organic amendments with or without the application of chemical fertilizers (28%) had been applied. However, in plots where stover alone was returned, the SOC declined from 2.0 to 1.3 but where stover was returned in combination with fertilizers, the loss was slightly higher at 39% (Kibunja 2007). On the average, the change of soil organic carbon was significantly influenced by manure addition ($p=0.01$), residue incorporation ($p=0.02$) and also declined significantly with time ($p=0.01$). However, the influence of the treatments and their combinations over time was not statistically significant. Measured soil carbon stocks over a period of 20 years indicate a gradual decline regardless of treatment applied from an average of 35 t C ha^{-1} to about 25 t C ha^{-1} as shown in Fig. 4.5a and b (Kamoni et al. 2007). Simulation modeling using ROTH-C-26.3 model supported the measured decreases in soil organic carbon and the model further predicted a lengthy and difficult amelioration process for such a degraded soil, (Ayaga 2000). Treatments with integrated chemical fertilizers, farmyard manure and crop residues were better at maintaining soil carbon stocks than chemical fertilizers or addition of crop residues alone.

However, the major research question arising from this trial still remains: what is the optimal quantity of organic inputs required to restore SOC at its original level (20 kg^{-1} SOC)? The inability to sustain SOC may be attributed to enhanced or higher rates of decomposition of soil organic matter compared to inputs under continuous cultivation. Given that both low quality organics (stover) and moderate quality organics (FYM) were applied, this may imply that an opportunity still exists for investigating the effect of the quality and quantity of organic materials on nutrient availability and plant uptake dynamics. Results arising from this researcher-managed trial were corroborated by another study that evaluated farmer experiences in the surrounding Kiambu district. Murage (1998) found that farmers use various combinations of manure, compost, refuse, crop residues and inorganic fertilizers to manage the soil fertility of their land. The most productive farms had high SOC content and had received high inputs while the non-productive farms had lower SOC contents and received correspondingly less inputs.

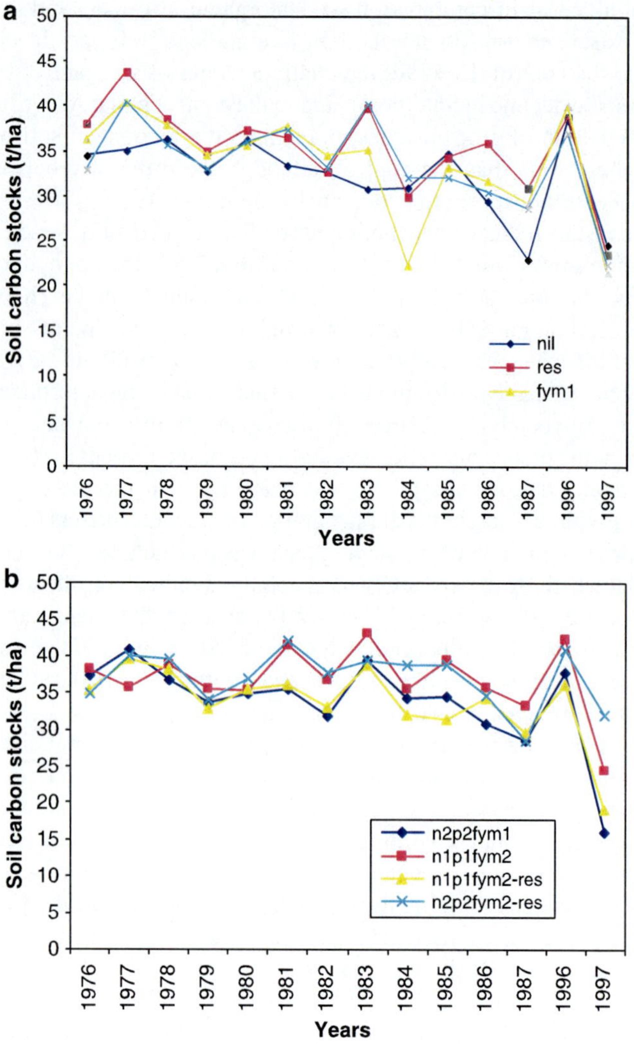

Fig. 4.5 Measured soil carbon stocks at Kabete long term experiment (Source: Kamoni et al. 2007). Legend: Nil – control; N1P1 – chemical fertilizer at 60 kg N ha^{-1} ± 26 kg P ha^{-1}; N2P2 – 120 kg N ha^{-1} ± 52 kg P ha^{-1}; FYM1 – farmyard manure at 5 t ha^{-1}; FYM2 – farmyard manure at 10 t ha^{-1}; Res – all harvested stover returned

A biometric analysis was carried out to determine which treatments contributed most to maize profitability over a period of 24 years of continuous cropping (Thuranira 2007). Models were fitted to establish the optimal levels of farm yard manure and organic fertilizer application. The results indicated that applying inorganic fertilizers together with farm yard manure was beneficial ($p = 0.000$) to the maize crop. Residues alone have no significant benefits to the crop ($p = 0.240$).

This is in agreement with the earlier work of Kapkiyai et al. (1998) who stated that recycling nutrients through livestock was more efficient that applying crop residues to the field. The study established that the optimal application level for farm yard manure is about 13 t/ha and that for inorganic fertilizers was 220 NP kg/ha (Thuranira 2007). These results indicate that higher inputs than are currently being applied at the Kabete long-term trial may be needed to restore the SOC to its former level of 2% at the beginning of the experiment in 1976.

Continuous cultivation with either sole NP fertilizers or without any inputs caused the greatest reduction in SOC of 53.5% and 37%, respectively. Application of fertilizer, farmyard manure and crop residues sustained the highest amount of SOC of 1.44%. Combined analysis over a 25-year period further showed that while SOC declined significantly ($p=0.01$) with time, the change in SOC was significantly influenced by manure addition ($p=0.01$) and residue incorporation ($p=0.02$). Continuous cultivation without organic inputs probably led to faster soil organic matter (SOM) turnover due to repeated tillage accompanied by increased aeration and reduced soil cover (Giller et al. 1997).

Data from this experiment were (re-)interpreted using simple models to find causes of success or failure with regard to sustained soil productivity (Janssen 2011). A two-pool model calculated the development of soil organic matter, and a practical equation estimated the residual effect of fertilizer P. Relative mineralization rate was 4% and 8% year^{-1} for original and newly formed soil organic carbon (SOC). Maize yielded 0.25 and 1.1 t ha^{-1} per g kg^{-1} of original and new SOC, respectively. Yields of fertilized maize increased initially as a result of increasing residual effects of applied P, but decreased later presumably because SOC declined to below a critical level of 16 g kg^{-1}. To maintain SOC above this level, about 10 tons of farmyard manure (dry matter) must be applied annually. Agronomic nutrient use efficiencies for fertilizer N and P were low, but the residual effect of P was high. The simple model outlined half a century ago adequately calculated build-up of new soil organic matter while the estimated residual effect of fertilizer P explained increasing crop responses to repeated P applications. However, the absence of data on nutrient uptake by the crop strongly limited the understanding of the experimental results.

Soil Total Nitrogen

Little change in total N stocks was recorded in the treatments with integrated fertilizer, manure and crop residues compared to the initial soil N content (Table 4.3). However, combined analysis indicated that total % N was significantly influenced by all the treatments applied. The no-input control plots had higher total N content compared to the plots with either NP or FYM fertilizers. This probably implies that application of inputs could have led to increased crop yields with resultant higher N uptake. On the other hand, the low crop yields in the control plots possibly took up comparatively less soil N than the fertilized treatments. Combined analysis indicated that N was significantly influenced by all the treatments applied, that is, farmyard manure ($p<0.01\%$), NP fertilizers ($p=0.22\%$) and crop residues ($p<0.01\%$). The decline of N was also highly significant ($p=0.01\%$) with time. However, soil organic carbon declined at a faster rate than total nitrogen. Use of NP fertilizers alone led to faster decline in % SOC (54%) and % total N (25%). Other results showed that soil pH in plots where no inputs were applied dropped from 5.5 to 4.2 over the same period. Soil bulk density increased slightly from 1.04 to 1.08 g cm^{-3} but not significantly (Table 4.3). Repeated use of chemical fertilizers alone may have led to increased soil acidification (Giller et al. 1997) and declining SOM (Palm et al. 1997) which may have hindered root development and nutrient availability.

Nitrogen Dynamics

The most prevalent form of mineral N was in the form of NO_3^--N and not NH_4^+-N throughout a 2 year study period from 2000 to 2001 (Kibunja 2007). A substantial amount of pre-season NO_3^--N ranging from 10.5 to 23.3 mg N kg was found in the plough layer (20 cm) of the soil (Fig. 4.6a). The concentration of NO_3^--N within the first 1 m soil level decreased with depth (4.5–5.8 mg N kg) as the crop matured, particularly at time of flowering (12 weeks after planting) as shown in Fig. 4.6b. This could be attributed to plant uptake and loss through leaching. A bulge of higher concentrations (13.6–29.3 mg N kg) was found in the lower levels from 1 to 3 m probably due to leaching. At harvest, there was an increase of NO_3^--N (11.5–14.1 mg N kg) in the top 1 m and a greater bulge (39.4–60.3 mg N kg) below the 1 m zone probably due to further leaching of both fertilizer-N and mineralized N (Fig. 4.6c). The amount of nitrogen found beyond the maize roots (100 cm) increased with time and reached a maximum at time of crop maturity. This could be attributed to more leaching of available N which coincided with the stage of decreased plant nutrient uptake. During the bean-growing season, a similar trend was noted. However, there was more mineral N within the top 1 m of soil during the bean-growing season than during the maize season. The bean crop had shallower roots and also fixed atmospheric N and may thus have spared the soil N and/or residual fertilizer N. This may have led to higher rates of mineral N loss through leaching. There was no significant difference ($p>0.05$) between the treatments with either organic or inorganic inputs except at the flowering stage (Fig. 4.5b). These results indicate that leaching is a significant open end to the nitrogen cycle leading to potential pollution of underground water. Similar work by Mekonnen et al. (1997) reported that the amount of leached nitrate-N was affected by the land use system e.g. more nitrogen was found at 0.3–1.5 m depth under maize than under either a natural weed fallow or a planted *Sesbania sesban* fallow. Sub-soil N (0.5–2.0 m) remained largely unchanged after three crops of unfertilized maize but decreased by 75–125 kg N ha^{-1} $year^{-1}$ under both natural and sesbania fallows (Hatermink et al. 1996). According to Mugendi et al. (2003), the agroforestry trees, *Leucaena leucocephala* and *Calliandra calothyrsus* intercepted leached N in the 100–300 cm depth through their deep roots.

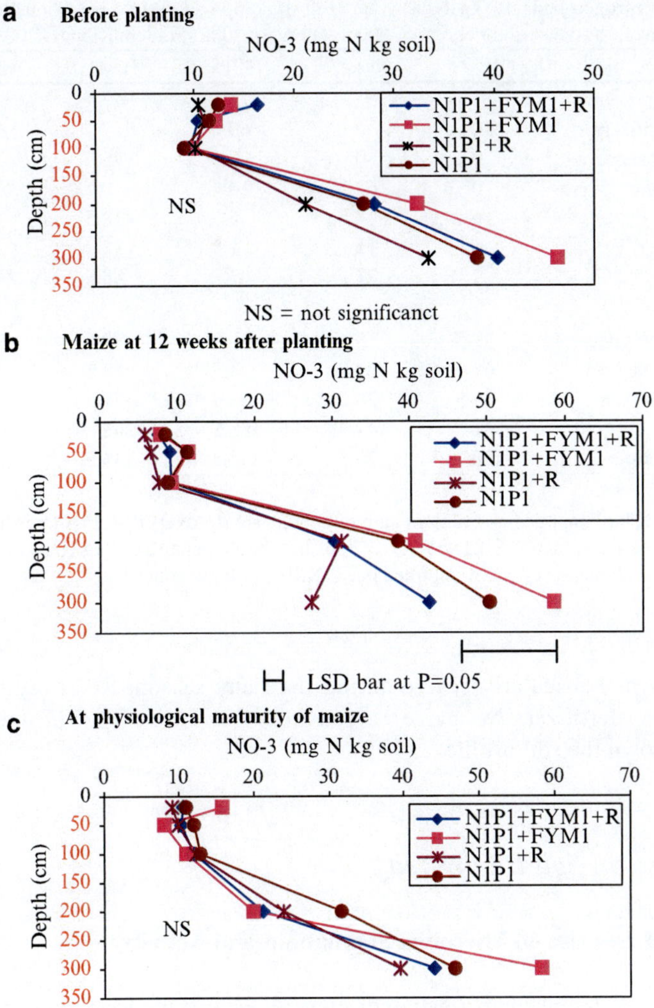

Fig. 4.6 Distribution of mineral N within a 0–3 m soil profile over time during the long rains maize season (Mar–Sept 2000), Kabete, Kenya. Legend: N1P1 – chemical fertilizer at 60 kg N ha^{-1} ± 26 kg P ha^{-1}; FYM1 – farmyard manure at 5 t ha^{-1}; R – all harvested stover returned (Source: Kibunja 2007)

Soil N Balances

Soil N balances showed that addition of farmyard manure, chemical fertilizers and crop residues contributed substantial amount of total N inputs to the soil (67–172 kg N ha^{-1} $year^{-1}$). However, only the treatment with integrated chemical fertilizers, manure and crop residues resulted in a positive N balance (39 kg N ha^{-1}). Harvested products (grains and stovers) and leached N in the top first meter of soil formed the main exports out of the system (Table 4.4). The loss of fertilizer

Table 4.4 Nitrogen budgets (kg N ha^{-1} $year^{-1}$) in the crop-soil system in various treatments at Kabete, Kenya, for the period February 2000 to February 2002 (Source: Kibunja 2007)

Treatments	NIL	Res	N1P1	FYM1	N1P1FYM1R
In					
Mineral fertilizers	0	0	60	0	60
Farmyard manure	0	0	0	75	75
Crop residues	0	62.5	0	0	91
N fixation	6	7.5	6	12.5	17
Seed N	1	1	1	1	1
Total In	**7**	**71**	**67**	**88.5**	**244**
Out					
Harvested products	49.5	55.5	77.5	79.5	76.5
Crop residues	68	62.5	74.5	88.5	91
Leaching (100 cm)	39.5	26.5	33.5	37	37.5
Total Out	**157**	**144.5**	**185.5**	**205**	**205**
Total N Balance	**–150**	**–73.5**	**–118.5**	**–116.5**	**39**

Legend
N_1 – 60 kg N ha^{-1} {applied as Calcium ammonium nitrate (CAN)}; P_1 – 26.4 kg P ha^{-1} {applied as Triple superphosphate (TSP)}; FYM_1 – 5 tons ha^{-1} boma manure (15 kg N and 4 kg P t^{-1} dry manure); R – all harvested stover returned back; NIL – no-input control

nitrogen was particularly severe during the rainy season. Recovery of applied nitrogenous fertilizers by maize ranged from 10% to 55% while the rest was leached down the soil profile.

Land Use on Soil Biodiversity

Effect of Land Use on Microbial Populations and Activity

The effect of continued application of mineral and organic fertilizers on soil agro-properties and soil microbial populations and activity, was studied in the Kabete long-term field experiment (Kibunja 2007; Kibunja et al. 2010). The study showed that use of manure alone or in combination with chemical fertilizers led to higher numbers of microbes and enhanced microbial respiration than use of chemical fertilizers alone (Fig. 4.7). The topsoil layer had significantly ($p=0.05$) higher microbial activity than the sub-soil regardless of management strategy. The study further showed that bacteria were more numerous {1 × 105 colonies forming units (cfu) g dry wt. soil −1} than fungi (1 × 103 cfu g dry wt. soil −1), which may lead to more soil organic matter (SOM) mineralization and less SOM retention in this cropping system. Integrated use of organic inputs such as farmyard manure and chemical fertilizers is recommended to maintain soil productivity under continuous cultivation.

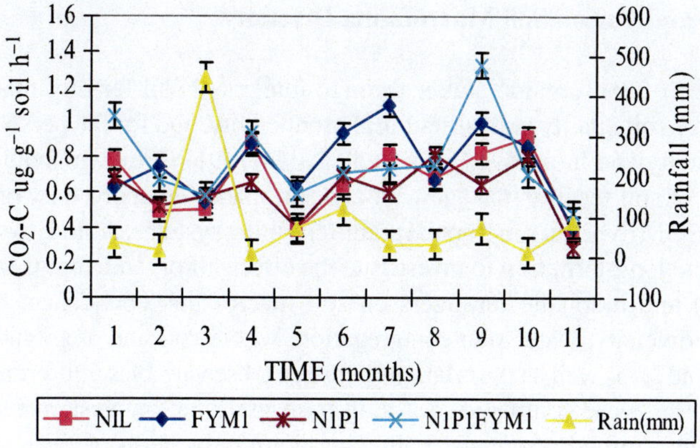

Fig. 4.7 Changes in the carbon dioxide (CO_2 – C μg g^{-1} soil h^{-1}) evolution upon application of various organic and inorganic inputs over time at the long term trial, Kabete, Kenya. Symbols: N – 60 kg N ha^{-1} {(as Calcium ammonium nitrate (CAN)}; P – 26.4 kg P ha^{-1} as triple superphosphate (TSP); FYM – 5 tons ha^{-1} manure (with 15 kg N and 4 kg P t^{-1} dry manure); R – all harvested stover returned back; *NIL* – no-input control; Rainfall in mm. *Bars* indicate the SE (Source: Kibunja 2007)

Effect of Land Use on Microbial Communities

The effects of crop manure and inorganic fertilizers on composition of microbial communities of central highland soils of Kenya are poorly known. A study was carried out on the 32 year old long-term trial at Kabete, Kenya to investigate the effect of land use of soil below ground biodiversity. The soil in the study site was treated with organic (maize stover (MS) at 10 tha^{-1}, farmyard manure (FYM) at 10 tha^{-1}) and inorganic fertilizers 120 kg N, 52.8 kg P (N2P2), N2P2 + MS, N2P2 + FYM, a control (C) and a fallow (F) for over 30 years. The 16S rRNA gene and 28S rRNA gene fingerprints of bacterial and fungal diversity were examined by PCR amplification and denaturing gradient gel electrophoresis (PCR-DGGE) separation, respectively (Kamaa et al. 2011). The PCR bacterial community structure and diversity was negatively affected by N2P2 and was more closely related to the bacterial structure in the soils without any addition (control) than that of soils with a combination of inorganic and organic or inorganic fertilizers alone. The effect on the composition fungal diversity by N2P2 was different than the effect on the composition of bacterial diversity since the fungal diversity was similar to that of the N2P2 + FYM and N2P2 + MS treated. However, soils treated with organic inputs clustered away from soils amended with inorganic inputs. Organic inputs had a positive effect on both bacterial and fungal diversity with or without chemical fertilizers. Results from this study suggested that total diversity of bacterial and fungal communities was closely related to agro-ecosystem management practices and may partially explain the yield differences observed between the different treatments (Fig. 4.7).

Effect of Land Use on Soil Macrofauna Diversity

Maintenance of soil organic matter through integrated soil fertility management is important for soil quality and agricultural productivity, and for the persistence of soil faunal diversity and biomass (Ayuke et al. 2011). Little is known about the interactive effects of soil fertility management and soil macrofauna diversity on soil aggregation and SOM dynamics in tropical arable cropping systems. A study was conducted in the Kabete long-term trial to investigate the effects of organic inputs (maize stover or manure) and inorganic fertilizers on soil macrofauna abundance, biomass and taxonomic diversity, water stable aggregation, whole soil and aggregate-associated organic C and N, as well as the relations between these variables. Differently managed arable systems were compared to a long-term green fallow system representing a relatively undisturbed reference. Fallowing, and application of farm yard manure (FYM) in combination with fertilizer, significantly enhanced earthworm diversity and biomass as well as aggregate stability and C and N pools in the top 15 cm of the soil. Earthworm abundance significantly negatively correlated with the percentage of total macroaggregates and microaggregates within macroaggregates, but all earthworm parameters positively correlated with whole soil and aggregate associated C and N, unlike termite parameters. Factor analysis showed that 35.3% of the total sample variation in aggregation and C and N in total soil and aggregate fractions was explained by earthworm parameters, and 25.5% by termite parameters. Multiple regression analysis confirmed this outcome. The negative correlation between earthworm abundance and total macroaggregates and microaggregates within macroaggregate could be linked to the presence of high numbers of *Nematogenia lacuum* in the arable treatments without organic amendments, an endogeic species that feeds on excrements of other larger epigeic worms and produces small excrements. Under the conditions studied, differences in earthworm abundance, biomass and diversity were more important drivers of management-induced changes in aggregate stability and soil C and N pools than differences in termite populations.

Conclusions

It was noted that none of the land management strategies applied so far at this trial have managed to sustain soil organic carbon at the original level. There is therefore need to improve carbon sequestration and reduce pathways of nutrient loss. Continuous cropping led to soil acidification especially where inorganic fertilizers were applied without any organic inputs. The Kabete soil is clayey in nature and is apparently resilient with a high capacity to supply N through mineralization but better management strategies are needed to improve nutrient utilization for crop growth. This can be done through use of deep-rooted shrubs/trees for sub-soil N capture; short-term immobilization of pre-season N through application of high C: N ratio materials such as stovers; use of cover crops to capture plant-available N between crop removal and planting of the next crop. Liming may be an important strategy to deal with rising soil acidification especially where inorganic fertilizers alone have been continuously applied without the addition of any organic inputs.

Way Forward

Several strategies, i.e., integrated use of organic and inorganic sources of inputs to build up soil organic matter, application of lime to address soil acidity, rotation of maize with deep rooted plants for nitrogen capture have been developed for each constraint and are awaiting validation. Using Geographic Information Systems (GIS) and satellite imagery tools, these results can be adapted and up-scaled to other agro-ecological zones in the country where small-scale farmers practise continuous cropping. However, it is pertinent to note that there is still a lot more awaiting to be investigated. For example, no substantive data on the impact of land management on crop-water dynamics and soil physical characteristics has been collected to-date. Secondly, no studies have yet been conducted on the dynamics of other major nutrients such as phosphorus, potassium, sulphur etc. and minor nutrients especially zinc and iron which are gaining international interests in human nutrition. Some work on the effect of land use on macro and micro biodiversity has now been covered but more understanding on pest and insect damage control is needed. There is further need to determine the quality and quantity of inputs required to revert the lost soil organic carbon back to its original state and also determine other limiting nutrients limiting crop productivity in this system. Efforts to validate and promote technologies from these studies should be enhanced to bring solutions to the small-scale farmers who are the end-users of this information.

Acknowledgements The authors wish to thank the Director, Kenya Agricultural Research Institute for permission to publish this work, the TSBF-AFNET Programme and European Union for financial support towards field maintenance and sample analyses.

References

Ayaga GO (2000) Improving the efficiency of phosphate utilization in low-input maize production in Kenya. PhD thesis. University of Nottingham, UK

Ayuke FO, Brussaard L, Vanlauwe B, Six J, Lelei DK, Kibunja CN, Pulleman MM (2011) Soil fertility management: impacts on soil macrofauna, soil aggregation and soil organic matter allocation. Appl Soil Ecol 48(2011):53–62

Bationo A, Waswa SB, Adamu A, Bado V, Moussa B, Iwaufor E, Kibunja C, Kihara J, Mucheru-Muna M, Mugendi J, Mwale C, Okeyo JM, Andren O, Roing K, Sedogo M, Tabo R (2007) Lessons learnt form long-term experiments in Africa. Symposium on success stories of agricultural long-term experiments, Sept 2007, Royal Swedish Academy of Agricultural and Forestry, KSLA, 28–29th May 2007, pp 30–35

Bekunda MA, Bationo A, Ssali H (1997) Soil fertility management in Africa. A review of selected research trials. In: Buresh RJ (ed) Replenishing soil fertility in Africa, vol 51, SSSA Spec. Publ. SSSA, Madison, pp 63–79

Birch HF (1958) The effect of soil drying on humus decomposition and nitrogen availability. Plant Soil 10:9–31

Chui JN (1994) The effects of nitrogen fertilization, bean residue and maize-bean cropping systems on nutrient uptake, yields and soil chemical properties in a semi-arid area. PhD thesis. University of Nairobi

FAO (1995) Planning for sustainable use of land resources: towards a new approach. In: Sombroek WG, Sims D (eds) Land and water bulletin, vol 2. FAO, Rome

Genstat 5 Committee (1993) Genstat 5 release 3 reference manual. Oxford University Press, Oxford

Giller KE, Beare MH, Lavelle P, Izac A-MN, Swift MJ (1997) Agricultural intensification, soil biodiversity and agroecosystem function. Appl Soil Ecol 6:1 (Special Issue: Agricultural Intensification, Soil Biodiversity and Agroecosystem Function in the Tropics), 3–16

Haggar JP, Tanner EVJ, Beer JW, Kass DCL (1993) Nitrogen dynamics of tropical agroforestry and annual cropping systems. Soil Biol Biochem 25:1363–1378

Harris GH, Hesterman OB, Paul EA, Peters SE, Janke RR (1994) Fate of legume and fertilizer ^{15}N in a long term cropping systems experiment. Agron J 86:910–915

Hartemink AE, Buresh RJ, Jama B, Janssen BH (1996) Soil nitrate and water dynamics in sesbania fallows, weed fallows and maize. Soil Sci Soc Am J 60:568–574

Heerink N (2005) Soil fertility decline and economic policy reform in sub-Saharan Africa. Land Use Policy 22:67–74

Hinga G, Muchena FN, Njihia CM (1980) Physical and chemical methods of soil analysis. Report, National Agricultural Research Laboratories, Nairobi, Kenya

Houba V, van Vark W, Walinga I, van der Lee JJ (1988) Plant analysis procedures, Part 7, Chapter 2.2. Department of Soil Science and Plant Analysis, Wagenigen, The Netherlands

IAEA (2001) Use of isotope and radiation methods in soil and water management and crop nutrition. Manual. IAEA -TCS –14. IAEA, Vienna

Jaetzold R, Schmidt H, Hornetz B, Shisanya C (2006) Farm management handbook of Kenya. Natural conditions and farm management, 2nd edn. Ministry of Agriculture/GTZ, Nairobi

Janssen BH (2011) Simple models and concepts as tools for the study of sustained soil productivity in long-term experiments. I. New soil organic matter and residual effect of P from fertilizers and farmyard manure in Kabete, Kenya. Plant Soil 339:3–16. doi:10.1007/s11104-010-0587-8

Jones E (1967) Soil fertility and crop rotation. East African specialist committee for soil fertility and crop nutrition. Muguga, Kenya, KARI

Jones E (1972) Principles for using fertilizers to improve red ferrallitic soils in Uganda. Exp Agric 8:315–332

Kamaa M, Mburu H, Blanchard E, Chibole L, Chotte JL, Kibunja C, Lesueur D (2011) Effects of organic and inorganic applications on soil bacterial and fungal microbial communities diversity and impacts of earthworms on microbial diversity in the Kabete Long-term trial, Kenya. Biol Fertil Soils 47:315–321. doi:10.1007/s00374-011-0539-3

Kamoni PT, Gicheru PT, Wokabi SM, Easter M, Milne E, Coleman K, Falloon P, Paustian K, Killian K, Kihanda FM (2007) Evaluation of two soil carbon models using two Kenyan long term experimental datasets. Agric Ecosyst Environ 122(2007):95–104

Kapkiyai JJ, Karanja NK, Woomer P, Qureshi JN (1998) Soil organic carbon fractions in a long-term experiment and the potential for their use as a diagnostic assay in highland farming systems of Central Kenya. Afr Crop Sci J 6(1):19–28

Kenya Agricultural Research Institute, (KARI) (1994) Fertilizer use recommendations project, vol 1–22. KARI, Nairobi

Kenya Soil Survey (KSS) (2001) Geographical distribution of humic nitisols in Kenya. Kenya soil survey. Kenya Agricultural Research Institute, Nairobi, Kenya

Keya SO (1977) Nodulation and nitrogen fixation in legumes in East Africa. In: Ayanaba A, Dart PJ (eds) Biological nitrogen fixation in farming systems of the tropics. John Wiley and Sons, pp 233–243

Kibunja CN (2007) Impact of long-term application of organic and inorganic nutrient sources in a maize-bean rotation to soil nitrogen dynamics and soil microbial populations and activity. PhD thesis. University of Nairobi, Nairobi

Kibunja CN, Gikonyo EW, Thuranira EG, Wamaitha J, Wamae DK, Nandwa SM (2003) Sustainability of long-term bean productivity using maize stover incorporation, residual fertilizers and manure. Paper presented at the 18th SSSEA Conference, 4–8 Dec 2000, Mombasa, Kenya, pp 217–222

Kibunja CN, Gikonyo EW (2000) Maintenance of soil fertility under continuous cropping in a maize-bean rotation (Long-term trial, NARL-Kabete). In: The biology and fertility of tropical soils. Report of the Tropical Soil Biology Fertility Programme. TSBF. 1997–1998. ISBN NO. 9966-9892- 4-2, pp 50–52

Kibunja CN, Mwaura FB, Mugendi DN (2010) Long-term land management effects on soil properties and microbial populations in a maize-bean rotation at Kabete, Kenya. Afr J Agric Res 5(2):108–113, 18 Jan 2010. http://www.academicjournals.org/AJAR. ISSN 1991-637X © 2010 Academic Journals

Mehlich A, Pinkerton A, Robertson W, Kempton R (1962) Mass analysis methods for soil fertility evaluation. Internal Publication, Ministry of Agriculture, Nairobi

Mekonnen K, Buresh RJ, Jama B (1997) Root and inorganic nitrogen distributions in sesbania fallow, natural fallow and maize fields. Plant Soil 188:319–327

Minae S, Akyeampong E (1988) Agroforestry potentials for the land use systems in the bimodal highlands of Eastern Africa, Kenya. Report. AFRENA No. 3: ICRAF Nairobi, 153 pp

Mugendi DN, Kanyi M, Irungu JB, Wamicha W, Mugwe JN (2003) Mineral-N movement in an agroforestry system in the Central Highlands of Kenya. In: Proceedings of the 18th SSSEA Conference, Mombasa, Kenya, 4–8 Dec 2000

Mughogho SK, Bationo A, Christianson CB, Vlek PLG (1986) Management of nitrogen fertilizers for tropical Africa. In: Mokwunye U, Vlek PLG (eds) Management of nitrogen and phosphorus fertilizers in sub-Saharan Africa. Martinus Nijhoff Publishers, Dordrecht, pp 117–172

Murage EW (1998) Soil carbon pools based on soil physical, chemical and biological fractionation and their relationship to soil fertility in central Kenya highlands. Msc thesis. University of Nairobi

Novozamsky I, Houba VJG, van Eck R, van Vark W (1983) A novel technique for multi-element analysis. Commun Soil Sci Plant Anal 14:239–249

Okalebo JR (2003) Management of inorganic resources for maintenance of soil productivity. In: Mureithi JG, Macharia PN, Gichuru M, Mburu M, Mugendi DN, Gachene CKK (eds) Proceedings of the 18th conference of the Soil Science Society of East Africa, Mombasa, Kenya, SSSEA

Onduru DD, De Jager A, Muchena FN, Gachimbi L, Gachini GN (2006) Socio-economic factors, soil fertility management and cropping practices in mixed farming systems of sub-Saharan Africa: a study in Kiambu, central highlands of Kenya. Int J Agric Res 2(5):426–439

Palm CA, Myers RJK, Nandwa SM (1997) Combined use of organic and inorganic nutrient sources for soil fertility replenishment. In: Buresh RJ (ed) Replenishing soil fertility in Africa, vol 51, SSSA Spec. Publ. SSSA, Madison, pp 193–217

Probert ME, Okalebo JR, Simpson JR, Jones RK (1992) The role of boma manure for improving soil fertility. In: Probert ME (ed) A search for strategies for sustainable dryland cropping in semi-arid eastern Kenya. Proceedings of a symposium held in Nairobi, Kenya 10–11 Dec 1990. ACIAR, Canberra, Australia, pp 63–69

Qureshi JN (1979) Annual report. Soils and Agricultural Chemistry Section, National Agricultural Laboratories, Nairobi, Ministry of Agriculture, Kenya

Qureshi JN (1991) The cumulative effects of N-P fertilizers, manure and crop residues on maize and bean yields and some chemical properties at Kabete. In: Proceedings of the 2nd KARI conference, Nairobi, Kenya 5–7 Sept 1991

Sanchez PA (1976) Properties and management of soils in the tropics. John Wiley and Sons, New York, 618 pp

Sanchez PA, A-M N, Izac IV, Pieri C (1996) Soil fertility replenishment in Africa: a concept note. In: Breth SA (ed) Achieving greater impact from research investments in Africa. Sasakawa Africa Assoc, Mexico City, pp 200–207

Sanchez PA, Shepherd KA, Soule MJ, Place FM, Buresh RJ, Izac A-MN, Mokwunye AU, Kwesiga FR, Ndiritu CG, Woomer PL (1997) Soil fertility replenishment in Africa: an investment in natural resource capital. In: Buresh RJ, Sanchez PA, Calhon F (eds) Replenishing soil fertility in Africa. SSSA Special Publication No 51. Soil Science Society of America & American Society of Agronomy, Madison, Wisconsin, USA pp 1–46

Scholes MC, Swift MJ, Heal OW, Sanchez PA, Ingram JSI, Dalal R (1994) Soil fertility research in response to the demand for sustainability. In: Woomer PL, Swift MJ (eds) The biological management of tropical soil fertility. Wiley, Chichester, pp 1–14

Siderius W, Muchena FN (1977) Soils and environmental conditions of agricultural research stations in Kenya, vol 5, Misc.Publication. Kenya Soil Survey, National Agricultural Laboratory, Nairobi

Simpson JR, Okalebo JR, Lubulwa G (1996) The problem of maintaining soil fertility in eastern Kenya: a review of relevant research, vol 41, ACIAR Monograph. Australian Centre for International Agricultural Research, Canberra, Australia, 60 pp

Stangel PJ (1995) Nutrient cycling and its importance in sustaining crop-livestock systems in sub-Saharan Africa: an overview. In: Powell JM, Fernandez-Riviera S, Williams TO, Renard C (eds) Livestock and sustainable nutrient cycling in mixed farming systems of sub-Saharan Africa, vol 2. Technical papers. Proceedings of an international conference, Addis Ababa, Ethiopia, 22–26 Nov 1993. ILCA, Addis Ababa, Ethiopia, pp 37–59

Stoorvogel JJ, Smaling EMA, Janssen BH (1993) Calculating soil nutrient balances in Africa at different scales. 1.Supra-national scale. Fertil Res 33:227–235

Strobel H (1987) Description of the first priority sites in the various districts. Fertilizer use recommendation project. Final report Annex III. Ministry of Agriculture, National Agricultural Research Laboratories, Nairobi, Kenya

Swift MJ, Seward PD, Frost PGG, Qureshi JN, Muchena FN (1994) Long-term experiments in Africa: developing a database for sustainable land use under global change. In: Leigh RA, Johnson AE (eds) Long-term experiments in agricultural and ecological sciences. CAB International, Wallingford, pp 229–251

Thuranira EG (2007) Analysis of a long-term maize yield field trial data to determine the effect of continuous application of fertilizers. University of Nairobi, July 2007

United Nations Educational, Scientific and Cultural Organization (UNESCO) (1974) FAO-UNESCO soil map of the world, vol VI. Africa. UNESCO, Paris, 299 pp

USDA-Soil Conservation Service (1975) Soil taxonomy. A system for soil classification. Soil survey staff. United States Department of Agriculture. SMSS Technical Monograph No. 19 Washington DC USA

Woomer PL, Muchena F (1996) Recognizing and overcoming soil constraints to crop production in tropical Africa. Afr J Crop Sci 4:503–518

Chapter 5
Management of Soil Fertility in a Long-Term Field Trial of Semi-arid Kenya

F.M. Kihanda and G.P. Warren

Abstract The long-term experiment was initiated in 1989 on a P-deficient soil in semi-arid Kenya where sustainability of the cereal/legume intercropping was assessed by monitoring trends in grain yield, soil organic C (SOC) and soil extractable P (Olsen method). Goat manure was applied annually at 0, 5 and 10 $t\,ha^{-1}$ and trends in grain yield were not identifiable because of season-to-season variations. SOC and Olsen P increased for the first 7 years of manure application and then remained constant. The residual effect of manure applied for 4 years only lasted another 7–8 years when assessed by yield, SOC and Olsen P. Mineral fertilizers provided the same annual rates of N and P as in 5 $t\,ha^{-1}$ manure and initially gave the same yield as manure, declining after 9 years to about 80%. There was a wide agreement between observed and predicted dry matter yields, soil labile P, but less for organic C using the Agricultural Production Simulator (APSIM) model. Soil nitrate was highest at the onset of the season and was highest where manure had been applied and was subject to loss through leaching or runoff

Keywords Manure • Crop yields • Soil properties • Semi-arid Kenya

Introduction

Mineral fertilizer and organic inputs are some of the key to improve on soil fertility and increase crop productivity. In semi-arid sub-Saharan Africa (SSA) it is usual to find great variability of rainfall and low accessibility to technical information and

F.M. Kihanda (✉)
Natural Resource Management, Kenya Agricultural Research Institute,
Embu Regional Research Centre, PO Box 27, Embu Kenya
e-mail: kihandafm@yahoo.com

G.P. Warren
Department of Soil Science, School of Human and Environmental Sciences,
The University of Reading, PO Box 233, Reading RG6 6DW, UK

A. Bationo et al. (eds.), *Lessons learned from Long-term Soil Fertility Management Experiments in Africa*, DOI 10.1007/978-94-007-2938-4_5,

markets, so the early and widespread adoption of fertilizers has not occurred. Manure application is one of the most effective ways of improving fertility in tropical soils. As an example, the production, distribution and application of manure is a vital part of sustainable smallholder arable farming around Kano in northern Nigeria, a semi-arid area with a high population and a long history of arable farming (Dennison 1961; Harris and Yusuf 2001). But organic nutrient sources have their own problems, such as limited supplies and the work of handling bulky materials. The effects and costs of organic and inorganic nutrient sources are different, but may be complementary. Thus the combined use of organic and inorganic nutrient sources is now considered a better way to maintain soil fertility, but comparative information on their long-term effects is scarce.

Both organic and inorganic sources have residual effects in the field. These effects are a vital component of sustainability because they smooth season-to-season variations in soil fertility and crop productivity, but they are difficult to assess quantitatively. Therefore, it is advantageous to undertake well-characterized medium-to-long term experiments rather than single-season trials, and to detail the interactions rather than averaging the responses over different seasons and environments (Tandon and Kanwar 1984). Long-term implies that primary objectives, treatments and management are not changed during the period under consideration, often regarded as at least 10 years, but unfortunately there is a paucity of long-term experiments in the semi-arid tropics (Laryea et al. 1995). However, long-term information on soil fertility can still be obtained by monitoring trends in the soil nutrient status even though the pattern of crops may change. The work described here is a development of work at one site of the multi-site experiment first reported by Gibberd (1995). In 1993, it was clear that manure at 5 $t\,ha^{-1}$ was widely beneficial, so advantage was taken of the significant soil fertility differences established by then (Warren et al. 1997b) to examine residual and long-term effects of soil fertility management.

The general objective of this work was to assess the sustainability of cereal/legume intercropping in semi-arid conditions, by assessing trends in crop yield and soil chemical data collected on long-term basis. Specific objectives were (i) to identify significant trends and differences in soil C and P status, (ii) to assess the direct and residual value of manure and (iii) to assess the relative effects of manure and fertilizer at the same rates of N and P addition and (iv) to identify the relative importance of manure and crop residues in maintaining soil organic matter (v) to determine the effect on soil nitrate (vi) In later years, to use the long-term dataset to test and validate agricultural models

Materials and Methods

Field Site

Table 5.1 summarizes the site characteristics of the trial site. The experimental site is located at Machang'a, Mbeere District, semi-arid areas of eastern Kenya (0°47′S, 37°40′E; 1,050 m altitude). It was cleared from native bush at the end of 1988,

Table 5.1 Site characteristics of the Machang'a trial site in the semi-arid eastern Kenya

Parameter characterized	Measurement
Latitude	0°47′S
Longitude	37°40′E
Altitude (m)	1067
Annual rainfall (mm)	774
pH (water)	6.55
pH ($CaCl_2$)	5.75
Exch. bases (cmol kg^{-1}) [Σ Na, Mg, K, Ca]	6.10
Organic C%	0.61
Total N%	0.06
Olsen P (mg kg^{-1})	0.94
Exchangeable K (mg kg^{-1})	331
Texture	Sandy clay loam
Class	Chromic cambisol

cropping was started in March 1989 and the manure treatments commenced in September 1989. The soil was a sandy clay loam containing 56.5%, 12.7% and 30.8% of sand, silt and clay respectively, with a pH of 6.55 (1:2.5 in water) and classified as a Chromic Cambisol, (Kenya Soil Survey, personal communication). Meteorological data were collected at the site. The biannual cropping seasons are identified by the month of peak rainfall and the rainfall for each season was assessed by summation of the total rainfall for October-January and March-June for the November and April seasons respectively.

Agronomy

The original experiment had nine treatments comprising three crop rotations and three fertility managements in factorial combination, each with three replicates laid out in randomized blocks. The crop rotations compared intercropping and two sole crop rotations. The fertility treatments were annual (a) additions of goat manure at 0, 5, and 10 t ha^{-1} a^{-1}. The results from 1989 to 1993 are given by Gibberd (1995). In February 1993, soil was sampled from all plots, and it was found that the different crop rotations had not created any significant differences in soil organic C (SOC), total N, extractable P (Olsen P) or exchangeable cations (Warren et al. 1997b). The intercropped rotation was continued without alteration to manure rates. The sole crop rotation with cereals planted in March and legumes planted in October was discontinued in order to create three new soil fertility treatments with intercropping as described below. The other sole crop rotation continued unchanged.

The soil fertility treatments are summarised in Table 5.2. Treatments C, A1 and A2 were maintained throughout from 1989 to date. The goat manure was obtained from the Ministry of Agriculture's Marimanti experiment station, Tharaka-Nithi District, in September each year, broadcast and immediately incorporated by hand

Table 5.2 Soil fertility treatments applied at the Machang'a trial. Manure was applied annually in September and fertilizer was applied every season

	Treatment	
Code	1989–1992	1993 todate
C	None	None
A1	5 t ha^{-1} a^{-1} manure	5 t ha^{-1} a^{-1} manure
A2	10 t ha^{-1} a^{-1} manure	10 t ha^{-1} a^{-1} manure
B1	5 t ha^{-1} a^{-1} manure	None (residual 1)
B2	10 t ha^{-1} a^{-1} manure	None (residual 2)
F	None	NPK fertilizer

digging. Treatments B1 and B2 assessed the residual effects after a final manure application in 1992 and treatment F assessed the effectiveness of mineral fertilizers on cropped and previously unfertilized soil. Manure was sampled and analysed for total C, N, P and K. In these years, the rates of fertilizer N and P in the new treatment F were adjusted in the April season so as to provide equal amounts of N and P in treatments F and A1. From 1997, the fertilizer treatment was N (51 kg ha^{-1}), P (12 kg ha^{-1}) and K (30 kg ha^{-1}) each season, providing approximately the same annual inputs of N and P as 5 t ha^{-1} of manure

The crops were (i) sorghum (*Sorghum bicolor,* cv. 954066), intercropped with cowpea (*Vigna unguiculata,* cv. M66); (ii) pearl millet (*Pennesitum typhoides,* cv. KPM1), intercropped with green gram (*Vigna radiata,* cv. N26) and (iii) maize (*Zea Mays,* cv. Katumani) intercropped with long duration pigeonpea (*Canjunus cajun*, cv. Kimbeere), the latter being a local variety. Cereals were planted in rows 70 cm apart at a spacing of 25 cm within rows, and the associated legume was planted at the same density in extra rows midway between the cereal rows. From 1989 to 1993, the treatments C, A1 and A2 carried a pattern of sorghum/cowpea for two seasons followed by millet/green gram for two seasons as described by Gibberd (1995). From October 1993 this was amended to a rotation (sorghum/cowpea sown in October and millet/gram sown in March) that closely follows typical local farming practice.

From October 1999, cropping was maize/pigeonpea. The first pigeonpea crop was sown in October 1999, grain harvested in May-August 2002 and the plots then cleared. The second pigeonpea crop was sown in October 2000 and harvests made in May-August 2001 and 2002 from the same plants. The intercropped maize was sown every October and March. The plots that were converted to soil treatments B1, B2 and F carried a rotation of sorghum; cowpea; millet; green gram, with cereals planted in March from 1989 to 1993. From October 1993 they carried the same rotation as treatments C, A1 and A2. Crop management included cultivation and weeding using locally available hand tools. At harvest, the grains and above-ground residues (leaves, stalks and threshing residues) were collected separately for each crop. They were air-dried and weighed at the site.

Soil Sampling and Analysis

Sampling of the soil commenced in February 1993 and was carried out at intervals of approximately 6 or 12 months, from the 0–20 cm horizon of all plots. Pits were dug

$30 \times 30 \times 20$ cm within steel frames driven into the soil, subsampled, air-dried and ground by hand to pass a 2-mm aperture sieve, as described by Warren et al. (1997b). Three sampling pits per plot were dug at the first (1993) and last (2002) sampling occasions and duplicate pits in 1997, but for reasons of economy, only a single sampling pit was used at the other occasions. Olsen P was measured colorimetrically after extraction for 30 min at 20°C and 1:20 w/v soil:reagent ratio with 0.5 M $NaHCO_3$ adjusted to pH 8.5. SOC was measured by heating finely-ground soil for 2 h at 130–135°C with $H_2SO_4/H_3PO_4/K_2Cr_2O_7$ mixture and back-titration with $(NH_4)_2.Fe(SO_4)_2$.

To determine nitrate N fresh soil (45 g) was weighed into a polypropylene bottle (250 ml), 1 M KCI (200 ml) added and shaken end-over-end (10 revolutions/minute) for 1 h on a rotary shaker. The suspension was allowed to incubae and consequent out on site, so the maximum time between sampling and extraction was about 3 h. Nitrate was measured with a "Heloflow" portable flow injection analyser (WPA Ltd., Ware, UK), using the Griess-Llosvay procedure (Keeney 1982). In this method, nitrate-N is reduced to nitrate by copperised Cd metal, reacted with NEDD (N- (1-naphthyl) ethylene diamine dihydrochloride) and measured colorimetrically (at 540 nm).

Statistical Methods

Season-to-season variability in the results makes it difficult to discern the differences and trends which may indicate the stability and sustainability of the cropping systems. Therefore the grain yield, Olsen P and SOC data were analysed statistically by analysis of variance for each season individually, and also by linear regression over sequences of seasons to search for trends. Statistical calculations were performed with INSTAT (Stern et al. 1990).

Regression modelling of grain yields

Empirical equations of the following form were fitted to grain yield (Y) for the continuous treatments (C, A1, A2):

$$Y = a + bR + cM^x + dR \times M^x \qquad (5.1)$$

where R was the seasonal rainfall (mm), M was the annual manure rate (t ha^{-1}), and *a*, *b*, *c*, *d* and *x* were fitted parameters. A preliminary matrix of correlation coefficients showed that Y was always correlated much more closely with $R \times M$ than either R or M alone. Then parameter *x* was selected to give the highest correlation between Y and $R \times M^x$, and the significant terms and regression coefficients were obtained using stepwise multiple regression.

Trends Over Time in Soil Properties Under Continuous Manure

For soil data, the trend over several seasons for each measurement and treatment was described by linear regression with time (T) in years as the explanatory variable.

Soil data for continuous manure treatment were considered in two phases, 1993–1997 and 1997–2002. 1997 was chosen as the dividing year because of the more intensive soil sampling carried out in that year. For 1989–1997, the value T=0 was set to 1 January 1989, approximately the start of the experiment, and for 1997–2002, T=0 was set to 1 January 1997. Comparisons between continuous manure treatments were then made by pooling the data for these treatments and a set of joint regression lines, either parallel to each other or diverging from a common origin at 1989, was fitted using multiple regression with factors (Draper and Smith 1981).

Assessment of Manure Residual Value

For each season from 1993, the residual value (RV) of manure for grain yield was assessed by the response to residual manure divided by the response to continuous manure, calculated as follows:

$$\mathrm{RV}i = \frac{\mathrm{Yield\ (B}i) - \mathrm{Yield\ (C)}}{\mathrm{Yield\ (A}i) - \mathrm{Yield\ (C)}} \tag{5.2}$$

where Ai denotes continuous manure at rate i , Bi denotes residual manure at rate i and C denotes the no manure treatment. RV should therefore vary between 1.0 for a residual effect as good as fresh manure and 0.0 when there is no residual effect. Residual values also were assessed by the responses in Olsen P and SOC, calculated by Eq. 5.2, with the substitution of yield by Olsen P or SOC values. The trends in yield and soil data were assessed by linear regression with time (T) in years as the explanatory variable. The value T=0 was set to 1 January 1993 since the new treatments commenced during 1993.

Comparison of Fertilizer and Manure

For each season from 1993 to 2002, the relative effect of fertilizer in comparison with manure was assessed by the fertilizer to manure ratio (FMR) for cereal grain yields calculated as follows:

$$\mathrm{FMR} = \frac{\mathrm{Yield\ (F)}}{\mathrm{Yield\ (A1)}} \tag{5.3}$$

FMR was also assessed for Olsen P and SOC, calculated by Eq. 5.3, with the substitution of yield by Olsen P or SOC values. The trends in yield and soil data were assessed by linear regression with time (T) in years as the explanatory variable. The value T=0 was set to 1 January 1993.

Simulations

The whole experimental period was simulated with an assumption that maize was planted every season (sowing dates 14th October and 18th March; 4 plants m^{-2}). Manure was applied and incorporated on 2nd October each year and fertilizer at sowing. The cultivations prior to sowing incorporated all crop residues. The simulations for each treatment were carried out as a single run, with all treatments initialized with identical inputs on 1 October 1989. Soil carbon in the surface 0–20 cm layer was based on the measured data in 1993 with an assumption that it declined in deeper layers (soil C deeper than 0.4 m was assumed not to mineralize). Similarly, soil labile-P was initialized using the measured Olsen-P data from the control treatment in 1993 and a factor of 2.5 used to convert Olsen-P (mg kg^{-1}) to labile-P (kg ha^{-1}).

The soil profile was sampled to determine bulk density and gravimetric soil water when dry (estimate of crop lower limit, LL) and also when wet (estimate drained upper limit, DUL). For the simulations shown below, smaller values of DUL have been assumed with the rooting depth set to 0.8 m resulting in Plant Available Water Capacity (PAWC) of 92 mm.

Results and Discussion

Weather

From 1990 to 2002, the mean annual rainfall was 789 mm, bimodally distributed with peaks in November and April (Fig. 5.1). Seasonal rainfall varied from 100 to 1,030 mm and appeared more variable from 1997 onwards (Fig. 5.1). Mean annual class "A" pan evaporation was 1993 mm, and the mean daily maximum and minimum temperatures were 29.0° and 16.6° respectively.

Continuous Manure

Grain Yields

For the unmanured plots (treatment C), the first season gave the highest yields of cereals and legumes (Fig. 5.2) and from then on, yields remained low every year. For the continuously manured plots (A1 and A2), yields were higher in the earlier years (approximately 1989–1995), when manure gave significant increases in cereal yield in almost every season and legume yield in three seasons. However, 10 t ha^{-1} manure did not give any significant extra grain yield compared to 5 t ha^{-1} manure (Fig. 5.2). In the period 1996–1998, all yields were low. For the cereals, this is

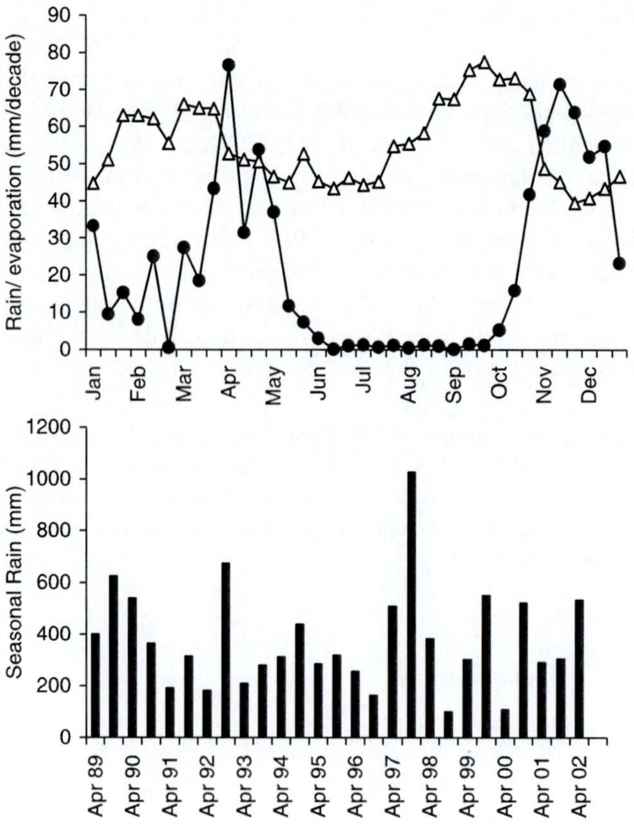

Fig. 5.1 Mean rainfall (•) and class 'A' pan evaporation (Δ) in "decades" (10-day periods), averaged over the period 1990–2002, and rainfall during each cropping season, assessed by summation of the rainfall from 1 October to the following 31 January for each November season and from 1 March to 30 June for each April season

attributed mainly to adverse weather conditions, *eg* the November seasons of 1996, 1997 and 1998 provided the highest and the two lowest rainfalls on record for that season (Fig. 5.1). Yields could be depressed by high rainfall.

In the November 1992 season, rainfall was the second highest on record but the sorghum yield was only about half that of the best season. In the extreme case of January and February 1998, an exceptionally extended rainy season caused complete loss of the November 1997 season's grain harvest, which rotted in the field and the yield had to be recorded as missing data. For the legumes, yields declined after 1992 (Fig. 5.2) and this was caused by an increasing incidence of disease. Cowpeas usually failed to set seed after 1995 because of root rot, which became endemic in the plots. After the intercropped plots were converted to the maize/pigeonpea system, grain yields were generally better (Fig. 5.2), although it is likely that this was in part due to more favourable rainfall patterns.

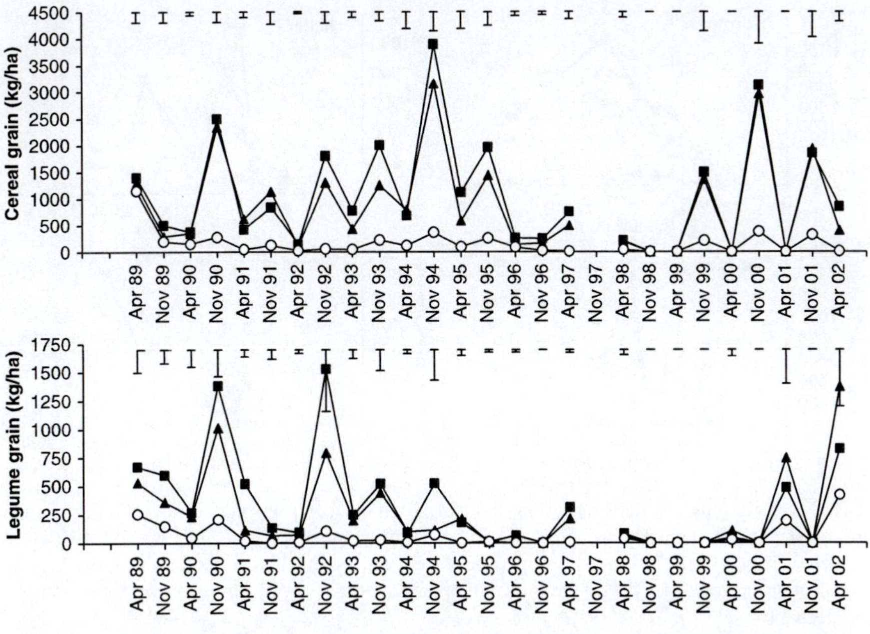

Fig. 5.2 Grain yields of cereals and legumes in each season with intercropping and soil fertility treatments C (•), A1 (▲) and A2 (■), in which 0, 5 and 10 $t\,ha^{-1}\,a^{-1}$ manure respectively were applied. *Vertical bars* equal the *s.e.d.* for each season

The relationships between grain yield and seasonal rainfall were plotted for each rate of manure (Fig. 5.3). For sorghum, it appeared that the yield was correlated with rainfall up to about 500 mm of seasonal rain, and that the response to extra rain was greater if manure was applied. Short duration sorghum in the tropics generally requires between 500 and 600 mm of well distributed rainfall to give optimum yields (Chantereau and Nicou 1994). An empirical regression equation gave a close description of the sorghum grain yield data for eight seasons in a period of 9 years, excluding (i) the initial season before manure was first applied (4/89); and (ii), the two seasons when rainfall exceeded 500 mm. The term for rainfall (R) in Eq. 5.1 was not significantly different from zero, and a very highly significant ($p<0.001$) relationship was found as follows:

$$\mathrm{Y} = a + b\mathrm{M}^{0.3} + c\mathrm{R} \times \mathrm{M}^{0.3} \quad (5.4)$$

where $a=146.7$, *s.e.* 78.9; $b=-867$, *s.e.* 92.7; $c=5.89$, *s.e.* 0.29 and $r^2=0.963$. This equation can be written in the following alternative form:

$$\mathrm{Y} = 146.7 + (5.89.\mathrm{R} - 867).\mathrm{M}^{0.3} \quad (5.5)$$

This suggests that (i) a certain minimum rainfall (R = 867/5.89 = 147 mm) is essential for grain production, which is otherwise negative according to this equation, and

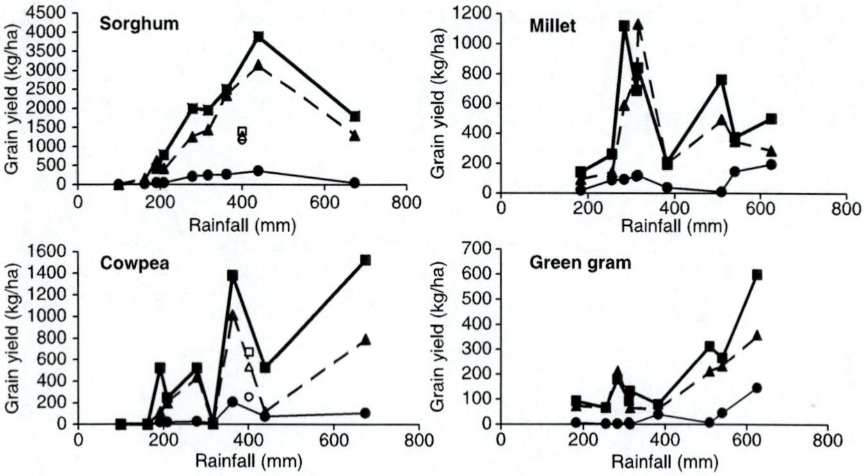

Fig. 5.3 Relationships between grain yield of sorghum, cowpea, millet and green gram, and seasonal rainfall, with intercropping and soil fertility treatments C (●), A1 (▲) and A2 (■), in which 0, 5 and 10 t ha^{-1} a^{-1} manure respectively were applied. Yields for the season before manure application started (April 1989) are shown in open symbols (○, Δ, □) for treatments C, A1 and A2 respectively

(ii) that above the minimum rainfall, yield depends on the product of R and M, showing that there is a strong positive interaction between rainfall and soil fertility.

For millet, cowpea and green gram, the yields in relation to rainfall were much more scattered (Fig. 5.3). For the legumes, it still appeared that the response to extra rain was better in the manured treatments. Close correlations of yield with M and R × M were not found, although for millet at rainfall less than 350 mm, the following relationship was significant:

$$Y = a + b\mathrm{M}^{0.1} + c\mathrm{R} \times \mathrm{M}^{0.1} \quad (5.6)$$

where $a = 85.8$, *s.e.* 92.2, $b = -994$, *s.e.* 313, $c = 5.18$, *s.e.* 1.11 and $r^2 = 0.774$.

Maize and pigeonpea yields were obtained in four and three seasons respectively, which were not enough data to obtain meaningful relationships between yield and rainfall.

Olsen P

Significant increases in Olsen P were caused by manure application. Compared to the control, treatment A2 caused significant increases in Olsen P at almost every sampling (Fig. 5.4), but the increase caused by treatment A1 was never large enough to be significant in any one season. Olsen P in treatment A2 was significantly more than in A1 in 9 out of 13 sampling occasions. In 1993, continuous

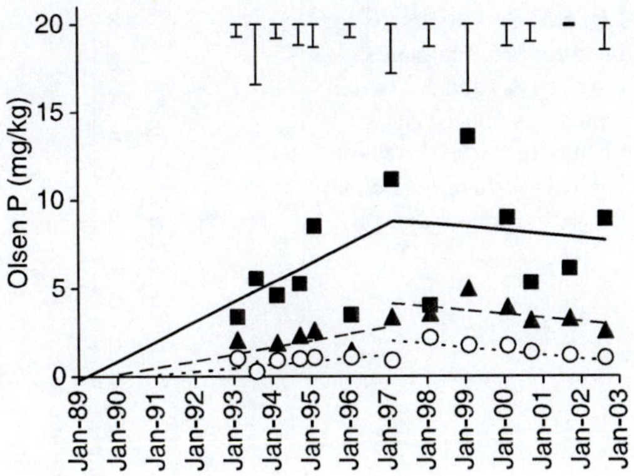

Fig. 5.4 Soil Olsen P for intercropped treatments C (○), A1 (▲) and A2 (■), in which 0, 5 and 10 $t\,ha^{-1}\,a^{-1}$ manure respectively were applied. *Vertical bars* equal the *s.e.d.* for each season

manure application had created significant differences between treatments C, A1 and A2 (Warren et al. 1997b), the Olsen P values being 1.0, 2.1 and 3.3 mg kg^{-1} respectively. By 2002, the Olsen P values were 1.0, 2.6, and 8.9 mg kg^{-1} respectively (Fig. 5.4). This suggested that trends in Olsen P should be observable.

From 1993 to 1997, the trends in Olsen P shown by regressions on time were small for treatment C, a little upward in A1 and distinctly upward in treatment A2 (Fig. 5.4). But, because of the variability of results, no trend line gradient for an individual treatment was significantly different from zero, although Olsen P must have been the same in all treatments in 1989 and a significant difference between treatments C and A2 had developed by 1993. When projected backwards to 1989, the fitted lines tended to a common origin as should be expected. Therefore, data for the three treatments were pooled and a set of joint regression lines was fitted, each of the following form, one equation for each treatment:

$$\text{Olsen P}\left(\text{mg kg}^{-1}\right) = a + b\text{T} \tag{5.7}$$

They started at a common origin with T=0 set to 1 January 1989, and had separate slopes. A good fit was obtained ($r^2=0.782$), shown in Fig. 5.4. The rate of increase of Olsen P was highly significant in treatment A2 but not in treatments C and A1, in agreement with the significant differences between treatments at each of the 1993–1997 samplings (Fig. 5.4).

From 1997 to 2002, the trends in Olsen P were slightly downward in all treatments (Fig. 5.4). The downward trends of individual treatments were not significantly different from zero or each other, but treatment effects in each season were consistently in the order C<A1<A2. These results were pooled and three parallel lines (each as Eq. 5.7) with separate intercepts were fitted by joint regression, with T=0 set to 1 January 1997. The joint downward slope was not significantly different from zero and Olsen P in treatment A2 was significantly more than in treatment A1 ($p=0.01$). The difference between treatments C and A1 was not quite large enough to be significant at the 5% level.

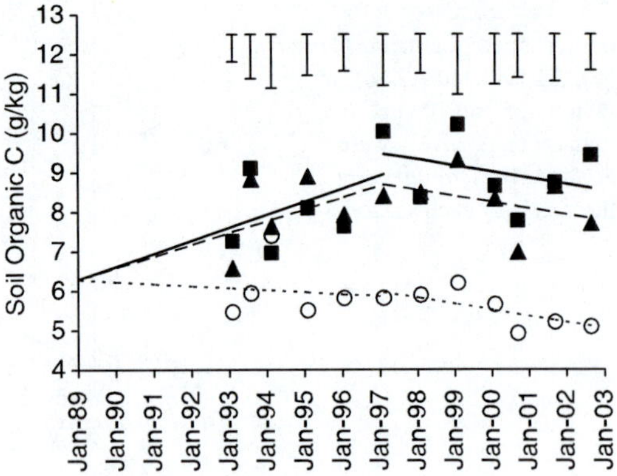

Fig. 5.5 Soil organic C for intercropped treatments C (○), A1 (▲) and A2 (■), in which 0, 5 and 10 t ha^{-1} a^{-1} manure respectively were applied. *Vertical bars* equal the *s.e.d.* for each season

Soil Organic C

Compared to the control, treatments A1 and A2 caused significant increases in soil organic C (SOC) at almost every sampling (Fig. 5.5), but the difference between treatments A1 and A2 was never significant.

From 1993 to 1997, the trend in SOC was small for treatment C and distinctly upward in treatments A1 and A2 (Fig. 5.5). But because of the variability of results, none of the gradients of the individual trend lines was significantly different from zero, although the initial SOC must have been the same in all treatments and a significant difference between treatments C and the manured treatments had developed by 1993. The initial SOC would be the same in all plots and when projected backwards in time, the trend lines tended to a common origin as might be expected. These data were pooled and set of joint regression lines based on a common origin and separate slopes were fitted to the equation:

$$\text{SOC}\left(\text{g kg}^{-1}\right) = a + b\text{T} \tag{5.8}$$

where $a=6.28$, *s.e.* 0.891, and $b=-0.053$, 0.296 and 0.328, for treatments C, A1 and A2 respectively, *s.e.* 0.155, ($F=8.98$; $r^2=0.658$). The rate of decline of SOC in the control treatment was not significant, while in the manured treatments, the rate of increase was not quite large enough to be significant and there was no significant difference between treatments A1 and A2. The loss of SOC through continuous cultivation have been previously reported in west Africa (Jones 1973; Jones and Wild 1975) while in Kenya it has been reported for the subhumid and semi-arid environments respectively (Swift et al. 1995; Kihanda et al. 2006) The data for treatments A1 and A2 were combined, and a regression line (Eq. 5.8) was estimated, where $a=6.28$, *s.e.* 0.865, and $b=0.312$, *s.e.* 0.145, in which the rate of increase of SOC was significant ($p=0.05$). These results are in agreement with the individual differences between treatments for the 1993–1997 samplings and show that annual manure application increased SOC up to 1997, but there was no difference between treatments A1 and A2.

From 1997 to 2002, the trends in SOC were slightly downward in all treatments (Fig. 5.5). The downward trends were not significantly different from zero or each other and treatment effects in each season were consistently in the order C < A2 ~ A1 (Fig. 5.5). These results were pooled, three parallel lines were fitted by regression, with T = 0 set to 1 January 1997. The set of regression lines was described by Eq. 5.8, where $a = 5.98$, 8.72 and 9.48 for treatments C, A1 and A2 respectively, *s.e.* 0.35, and $b = -0.154$, *s.e.* 0.081. ($F = 692$; $r^2 = 0.994$). The combined downward slope was still not significantly different from zero, SOC in treatments A1 and A2 was significantly more than in treatment C ($p = 0.001$), and there was no significant difference between treatments A1 and A2. In the current experiment, manure increased SOC by 52% and 85% in treatment A1 and A2 respectively compared with the control treatment. This is comparable to 40% increase caused by application of 4.9 $t\,ha^{-1}$ over 15 years in Nigeria (Bache and Heathcote 1969). In West Africa, De Ridder and van Keulen (1990) concluded that 5 $t\,ha^{-1}$ of manure was needed to maintain SOC. These results suggest that in the manure treatments after 1997, a new dynamic equilibrium had been reached between C inputs and decomposition.

Residual Manure Treatments

Grain Yields

The assessment of manure residues started in November 1993, and significant differences between continuous and residual manure treatments were not normally found in any one subsequent year because of the variability. In many years there were no significant effects of manure, either continuous or residual, on crop yields. Only cereal yield data were used because legumes were affected by disease. Residual Value (RV) could not be calculated for seasons of severe drought or missing data and the values were rather scattered (Fig. 5.6). Nevertheless, the linear regression between RV and time showed a highly significant ($p = 0.001$) downward trend, described by the following equation, where time (T) was zero for 1 January 1993.

$$\mathrm{RV} = a + b\mathrm{T} \tag{5.9}$$

where $a = 0.820$, *s.e.* 0.087, and $b = -0.0727$, *s.e.* 0.0157 ($r^2 = 0.495$). The confidence limits ($p < 0.05$) for the fitted mean regression line were plotted and projected back to April 1993 (Fig. 5.6). They showed that (i) from April 1993 to December 2001, RV was significantly less than one but more than zero, (ii) at April 1993, RV was not significantly different from one (this should be the case since the last manure application was the previous year), and (iii) by January 2002, RV was not significantly different from zero. These data suggest that the effects of manure lasted approximately 8 years, from the first residual effect season (November 1993) until December 2001. This is a longer period than has commonly been reported in semi-arid dryland agriculture. In Katumani Kenya, Ikombo (1984) reported a 3-year manure residual,

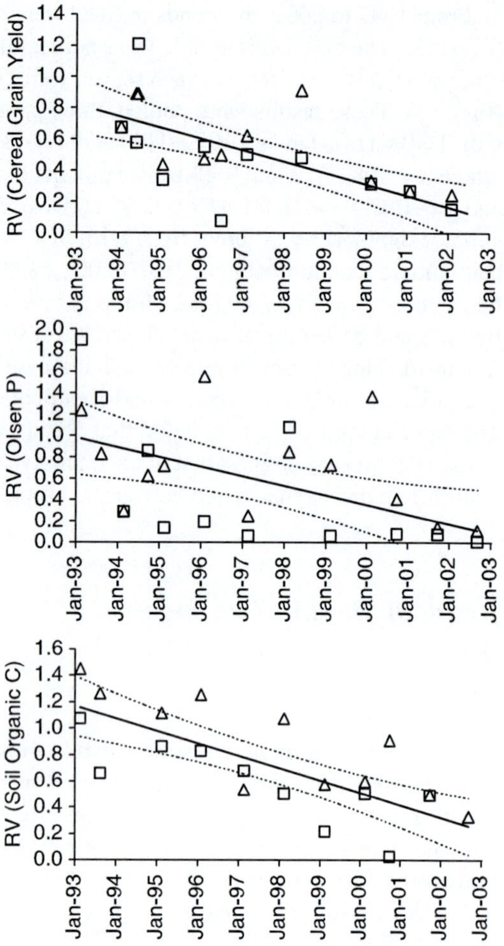

Fig. 5.6 Residual value (RV) of manure, calculated as response to residual manure (treatment B) divided by the response to continuous manure (treatment A), at 5 t ha^{-1} manure %) and 10 t ha^{-1} manure (□) for cereal grain yield, Olsen P and SOC. *Solid lines* show the fitted linear regression of RV with time (T years), starting 1 January 1993

while in India 2–3 seasons (Singh and Desai 1991) and two seasons in Botswana (Carter et al. 1992). Williams et al. (1995) showed that the annual breakdown for manure was in the ratio of 50:40:10 over 3 years.

Olsen P

Residual Value data were rather scattered (Fig. 5.6), but the linear regression of RV with time showed a highly significant ($p=0.01$) downward trend, described by Eq. 5.9, where $a=0.980$, *s.e.* 0.165, and $b=-0.0897$, *s.e.* 0.0305 ($r^2=0.265$). The confidence limits ($p<0.05$) for the fitted mean regression line were wider than for grain yield data (Fig. 5.6) and showed that by about September 2000, RV was not significantly different from zero

Soil Organic C

The Residual Value data showed a clear downward trend (Fig. 5.6), and the linear regression of RV with time was very highly significant ($p=0.001$), described by Eq. 5.9, where $a=1.169$, *s.e.* 0.107, and $b=-0.0941$, *s.e.* 0.0183 ($r^2=0.568$). The confidence limits ($p<0.05$) for the fitted regression line showed that by September 2002, RV had almost reached the point of being not significantly different from zero. These data suggest that the residual effects of manure on SOC lasted approximately 8 years, until 9 years after the final manure application.

Comparison of Fertilizer and Manure

Grain Yields

Only cereal yields were considered because legumes were badly affected by disease in the period under consideration. In every season, there was no significant difference between treatments F and A1 in grain yield. The relative effect of fertilizer in comparison with manure (FMR) showed a downward trend (Fig. 5.7), and the linear regression of FMR with time was significant ($p=0.05$). The confidence limits ($p<0.05$) for the fitted regression line showed that by the April 2002 season, FMR was significantly less than one. Therefore, by 2002, fertilizer was not as effective as manure even though the same amounts of N and P had been applied from 1993 to 2002.

Olsen P

During the period from 1993 to 2002, Olsen P was on average higher in treatment F (4.2 mg kg^{-1}) than in treatment A1 (3.0 mg kg^{-1}) even though the total amount of P applied since the start of the experiment was greater in treatment A1 than in treatment F, the quantities being approximately 261 and 159 kg ha^{-1} respectively. The slope of the regression of FMR with time was not significantly different from zero (Fig. 5.7), so fertilizer consistently increased Olsen P more than manure at the same rate of P application.

Soil Organic C

There was no significant trend in FMR for SOC with time over the period 1994–2002 (Fig. 5.7) although the apparent trend was up rather than down. In 1994, SOC was significantly less in treatment F than in treatment A1, as would be expected since treatment A1 had received five applications of manure at 5 tha^{-1}, while treatment F had received none. By 2002, FMR was not significantly different from 1.0 (Fig. 5.7). These results are tentative evidence that fertilizer had increased the soil organic matter (Fig. 5.7).

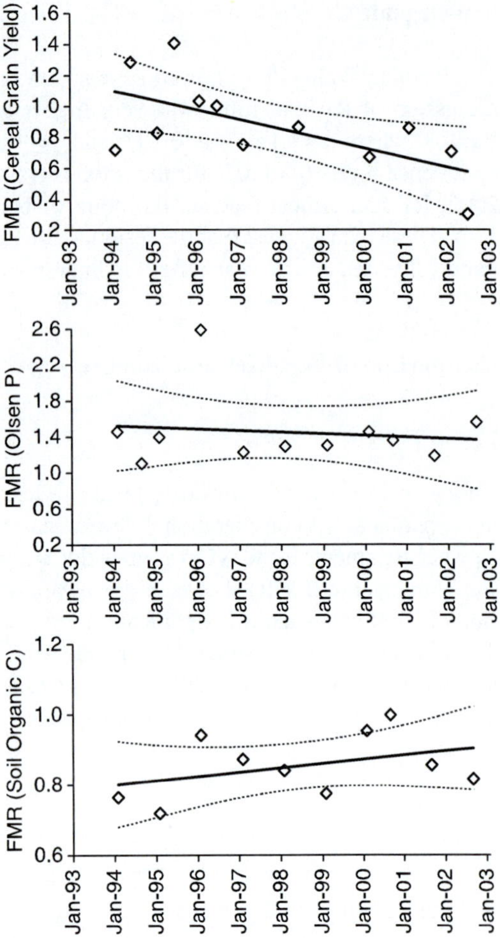

Fig. 5.7 Relative effect of fertilizer and manure (FMR; ◊), calculated as the ratio between the results for treatments F and A1 (fertilizer and 5 t ha^{-1} manure respectively) for cereal grain yield, Olsen P and SOC. *Solid lines* show the fitted linear regression of FMR with time (T years, starting 1 January 1993)

APSIM Model and Nitrate Variations

The data from a long-term experiment were also used to validate the performance of the Agricultural Production Simulator (APSIM) model in predicting the crop dry matter yield, soil phosphorus (soil-P) and soil organic carbon (SOC) on P deficient soils. The experiment was simulated using the APSIM P-aware maize module even though maize was not the test crop prior to the November 1999 season. The model performance was tested in terms of dry matter yields, soil P and organic carbon. There was a wide agreement between observed and predicted dry matter yields, soil labile P, but much less for organic C (Micheni et al. 2004).

Topsoil nitrate concentrations were highest at the start of the season and within about 15 days. In particular soil nitrate was normally higher at the beginning of the season in soils which had been manured recently and at the end of the season in soil with 10 t ha^{-1} applied recently. Because the high manure rate resulted in surplus

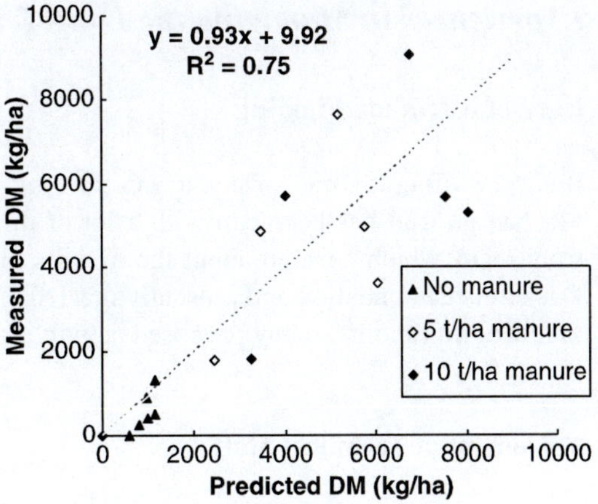

Fig. 5.8 Comparison of measured and predicted dry matter yields of maize for seasons from November 1999 using the APSIM model. The *dashed line* is the 1:1 relationship. The fitted linear regression equation and correlation coefficient are shown (Source: Micheni et al. 2004)

nitrate at the end of the season, a maximum manure rate of 5 $t ha^{-1}$ is suggested to reduce losses of N (Kihanda et al. 2004; Warren et al. 1997a).

Conclusions

- Machang'a manure experiment described here is one of few existing field experiments in semi-arid Africa of >10 years' duration with constant soil fertility treatments. It is representative of the farming systems of this region because it has used the most important grain crops of the region, a typical crop rotation, has components with manure and fertilizer, and was conducted in a "near-farm" situation with local management.
- Manure at 5 $t ha^{-1}$ increased SOC, but gave only a small increase in Olsen P, probably because P supply did not greatly exceed demand. Manure at 10 $t ha^{-1}$ gave no extra SOC over that generated by 5 $t ha^{-1}$ manure, and the extra C applied must have been lost. On the other hand, the higher manure rate increased Olsen P substantially because supply exceeded demand and there are few loss mechanisms for P.
- Results for the residual manure treatments showed that the residual effect of manure could last at least 7 years, a longer period than had been expected. Manure applications can be made intermittently and P and some N are stored in the soil. Nutrient supply in intermediate years may be improved with mineral fertilizers although they cannot maintain fertility in the long run research.
- Soil nitrate was highest at the onset of the season and was highest where manure had been applied and was subject to loss through leaching or runoff
- The data from a long-term experiment were also used to validate the performance of the Agricultural Production Simulator (APSIM) model in predicting the crop dry matter yield, soil phosphorus (soil-P) and soil organic carbon (SOC) on P deficient soils. There was a wide agreement between observed and predicted dry matter yields, soil labile P, but less for organic C.

Experiences in Managing the Long Term Experiment

Lack of Adequate Funding

It is very difficult attract grants to run the long-term Experiment. For example, the Machan'ga trial has been run with a lot of difficulties. Initially, the funding was from ODA which covered about the first six or so years of experimentation, then Rockefeller Foundation and presently the TSBF has been supporting. However due to low or no funding, many proposed activities are not accomplished.

Movement of Technical Staff

Some of the Technical staff managing the experiment are either transferred, retired, or move out to other institutions in pursuit of higher education. In some cases, others are laid off due to lack of funds. This bring lack of continuity in the trial management and often affects data quality.

Use/Misuse of Archived Samples and Datasets

During the period of experimentation, we are able to collect a lot of soil and plant samples that are archived. We are also in possession of huge database. Often there are many people (students, scientists etc.) who are only interested in using the data sets and sometimes they do not recognize acknowledge the people who are involved in running of the trial.

Acknowledgements We thank the following: The Rockefeller Foundation and TSBF for encouragement and financial support of this work; P.I. Mutwiri, C. Ndwiga and A. Kinyua for field site management; the Director, KARI Embu RRC for provision of facilities; and V. Gibberd and the Department for International Development, UK, for support of, and information about, the earlier years of the field work.

References

Bache BW, Heathcote RG (1969) Long term effects of fertilizers and manure on soil and leaves of cotton in Nigeria. Exp Agric 5:241–247

Carter DC, Harris D, Youngquist JB, Persaud N (1992) Soil properties, crop water use and cereal yields in Botswana after additions of mulch and manure. Field Crop Res 30:97–109

Chantereau J, Nicou R (1994) Sorghum. Macmillan, London

De Ridder N, van Keulen H (1990) Some aspects of the role of organic matter in sustainable intensified arable farming systems in the west African semi-arid tropics (SAT). Fertil Res 26:299–310

Dennison EB (1961) The value of farmyard manure in maintaining fertility in northern Nigeria. Emp J Exp Agric 29:330–336

Draper NR, Smith H (1981) Applied regression analysis, 2nd edn. Wiley, New York

Gibberd V (1995) Yield responses of food crops to animal manure in semi-arid Kenya. Trop Sci 35:418–426

Harris F, Yusuf MA (2001) Manure management by smallholder farmers in the Kano close-settled zone, Nigeria. Exp Agric 37:319–332

Ikombo BM (1984) Effects of farmyard manure and fertilizers on maize in semi-arid areas of Eastern Kenya. East Afr Agric For J 44:266–274

Jones MJ (1973) The organic matter content of soils of the savanna soils of west Africa. J Soil Sci 24:42–53

Jones MJ, Wild A (1975) Soils of the West African Savanna, vol 55, Technical Communication. Commonwealth Bureau of Soils, Harpenden

Keeney DR (1982) Nitrogen availability indices. In: Page AL, Millet RH, Keeney DR (eds) Methods of soil analysis, part II, vol 9, 2nd edn, Agronomy Monograph. American Society of Agronomy, Madison, Wi, USA

Kihanda FM, Warren GP, Atwal SS (2004) The influence of goat manure application on crop yield and soil nitrate variations in semi-arid eastern Kenya. In: Bationo A (ed) Managing nutrient cycles to sustain soil fertility in Sub-Saharan Africa. Academy Science Publishers in association with the Tropical Soil Biology and Fertility Institute of CIAT, Nairobi, pp 173–186

Kihanda FM, Warren GP, Micheni AN (2006) Effect of manure application on crop yield and soil chemical properties in a long-term trial of semi-arid Kenya. Nutr Cycl Agroecosyst 76:341–354

Laryea KB, Andres MM, Pathak P (1995) Long term experiments on alfisols and vertisols in the semi-arid tropics. In: Lal R, Stewart BA (eds) Soil management: experimental basis for sustainability and environmental quality, CRC Press, Boca Raton, pp 267–292

Micheni AN, Kihanda FM, Warren GP, Probert ME (2004) Testing the APSIM model with experimental data from the long-term experiment at Machanga (Embu), Kenya. In: Delve RJ, Probert ME (eds). Modelling nutrient management in tropical cropping systems. Australian Centre for International Research (ICIAR) proceedings No.114, Caberra, pp 110–117

Singh RP, Desai GM (1991) Farmers' fertilizer practices in the semi-arid tropics of India. Progress Report 100, Economics Group, Resource Management Program, ICRISAT. International Crops Research Institute for the Semi Arid Tropics, Patacheru

Stern RD, Knock J, Burn RW (1990) INSTAT. Statistical Services Centre, The University of Reading, Reading

Swift MJ, Seaward PD, Frost PGH, Qureshi JN, Muchena FN (1995) Long term experiments in Africa: developing a database for sustainable land use under global change. In: Leigh RA, Johnston AE (eds) Long-term experiments in agricultural and ecological sciences. CAB international, Wallingford, pp 229–251

Tandon HLS, Kanwar JS (1984) A review of fertilizer use research on sorghum in India, vol 8, ICRISAT Research Bulletin. International Crops Research Institute for the Semi Arid Tropics, Patancheru

Warren GP, Atwal SS, Irungu JW (1997a) Soil nitrate variations under grass, sorghum and bare fallow in semi-arid Kenya. Exp Agric 33:321–333

Warren GP, Muthamia J, Irungu JW (1997b) Soil fertility improvements under manuring in semi-arid Lower Embu and Tharaka-Nithi. In: Fungoh PO, Mbadi GCO (eds) Focus on agricultural research for sustainable development in a changing economic environment. Proceedings of the 5th KARI Scientific Conference, Nairobi, 14–16 Oct 1996. Kenya Agricultural Research Institute, Nairobi, pp 151–163

Williams TO, Powell JM, Fernandez-Rivera S (1995) Manure utilization, drought cycles and herd dynamics in the Sahel: implications for cropland productivity. In: Powell JM, Fernandez-Rivera, S, Williams O, Renard C (eds) Livestock and sustainable nutrient cycling in mixed farming systems of sub-saharan Africa, vol II: Technical Papers. Proceedings of an international conference held in Addis Ababa, Ethiopia, 22–26 Nov 1993. International Livestock Centre for Africa, Addis Ababa

Chapter 6
Long-Term Soil Fertility Trials in Niger, West Africa

A. Abdou, S. Koala, and A. Bationo

Abstract Soils of tropical Africa have been formed mainly from Pre-Cambrian materials and others from erosion process. Because of the type of parent material, the soils are characterized by poor native fertility surface crusting and low water holding capacity.

In Africa 65% of the agricultural land, 31% of the permanent pasture land, and 19% of the forest and woodland is affected by human-induced soil degradation. Nutrient depletion is the most important element in the land degradation equation.

With increasing population pressure on the land, the traditional farming system (natural fallow) to restore soil fertility is no longer possible. The present farming systems are unsustainable and destructive to the environment with negative plant nutrient balances resulting to mining of nutrients.

Many long term trials show that although application of mineral fertilizers is an effective mean of increasing yields in arable farming systems, mineral fertilizers alone cannot sustain yields in the long run. When mineral fertilizers are combined with crop residue or manure, added to cereal-legume intercrops or rotated, sustainable production can be obtained.

A. Abdou (✉)
Soil Fertility, TSBF-ICRISAT, BP: 12404 Niamey, Niger
e-mail: a.abdou@icrisatne.ne

S. Koala
Tropical Soil Biology and Fertility Institute of International Centre for Tropical Agriculture (CIAT-TSBF), ICIPE Duduville Campus off Kasarani Road, P.O. Box 823-00621, Nairobi, Kenya
e-mail: s.koala@cgiar.org

A. Bationo
Soil Health Program (Research & Extension), CSIR Office Complex #6 Agostino Neto Road Airport Residential Area, PMB KIA 114 Airport – Accra, Ghana
e-mail: abationo@agra-alliance.org

A. Bationo et al. (eds.), *Lessons learned from Long-term Soil Fertility Management Experiments in Africa*, DOI 10.1007/978-94-007-2938-4_6,

Any future interventions redressing distorted nutrient budgets, to be sustainable have to develop an integrated nutrient management, conceptualized as the judicious manipulation of the nutrient inputs and outputs processes.

Keywords Mineral fertilizer • Crop residue • Manure • Cropping system • Soil fertility • Long-term trial

Introduction

Soil degradation, loss of organic matter, low soil fertility and yields, poverty, high carbon dioxide (CO_2), global climatic changes are the main factors reducing crop production in the world. In the Sahel, low rainfall and its variability and distribution, dry spells and other adversities such as climatic factors affect crop production. Production losses are mainly due to drought (2/3) and cricket attack (1/3) (Nanga 2005). Water balance in the region is positive only 3 months of the year; meaning that water still a factor for crop production in a region where 90% of the population is rural and depend on subsistence rain-fed agriculture. Millet is the main crop in the Permanent Inter-state Committee for Drought Control in the Sahel (CILSS) countries with 45% of cereal production followed by sorghum (28%) and maize (11%). Niger is the second after Burkina Faso with 27% of cereal production. Niger, with a population of 12.94 million in 2006, is one of the food deficit countries in the world (CILSS/Agrhymet 2005). Only 12% of the country has an annual rainfall of 600 mm and 10% has 350–600 mm. A cereal crop needs at least 300 mm if well distributed (Moustapha 2003). Cereal production in 2004/2005 was estimated at 2,449,900 tones with a negative balance of 223,350 tones, equivalent to 7.5% of Niger population needs or 2,991,600 tones (Nanga 2005). Niger is the poorest country in the world according to the United Nations Development Program (UNDP) classification based on Human Development Index (HDI). Ninety eight percent of the cereal production in Niger is from rainwater while rice, which is the principal irrigated crop makes up less than 2% of the total cereal production. In 2004, rice production decreased to only 0.5% of cereal production (FAO 2004).

Population pressure has reduced cultivable area and traditional fallow is no longer feasible. It is widely known that millet is a crop well adapted to Sahel conditions but the combination of soil fertility constraints and low rainfall has reduced crop productivity. In this context of soil degradation and poor climatic conditions, farmer's practice needs to be adapted. The role of the International Center for Research in the Semi-Arid Tropics (ICRISAT) is to research and subsequently propose to farmers a combination of adaptable soil fertility technologies, water use-efficient methods and improved seed varieties that can significantly improve crop yields.

The purpose of this study is to assess how to improve natural resource management in poor soil and climatic conditions. Phosphorus (P), Nitrogen (N), manure interactions under erratic rainfall conditions are examined on a long term basis.

Experiments addressing several research themes were carried out along a bioclimatic gradient in several benchmark locations in West Africa using commonly developed research protocols. The long-term on-station trials from which technologies originated were tested on-farm with the objective of identifying sustainability indicators and optimizing the use of organic and inorganic resources available to the land users. Soil samples from the long-term trials were collected for measurement of these parameters.

Since 2000, these long-term on-station trials and others in different sites have been run under the Tropical Soil Biology Institute of CIAT (TSBF) Network in collaboration with ICRISAT with the aim of helping farmers and land managers combat soil nutrient depletion and land degradation, which are both serious threats to food production on the continent. The objective of these network activities is to develop and implement management options that both mitigate soil degradation, deforestation and biological resource losses and enhance local economies while protecting the natural resource base.

This chapter highlights the longest on-station trials run by TSBF Network in Niger, West Africa. The second part gives a brief description of other trials from different locations. The results of these experiments are reported every year in an annual report.

Long-Term Millet Trials Phosphorus, Nitrogen, Crop Residue (CR), Manure and Soil Tillage

Fields close to the village receiving high organic matter due to human and animal activities are more productive compared to those further from the homestead. Prudencio (1993) has observed such fertility gradient between fields closest to the homestead (home gardens/infields) and those furthest (bush fields/outfields). Soil organic carbon contents of between 11 and 22 g/kg have been observed in home gardens compared with 2–5 g/kg soil in bush fields. In a comparative study at Karabédji-Niger on degraded lands (bush fields) and non degraded (infields) it was observed that millet grain yield across years and fertilizer averaged only 800 kg/ha in bush fields and 1,360 kg/ha on infields. Recovery of fertilizer N (RFN) applied varied considerably and ranged from 17% to 23% on bush fields and from 34% to 37% on infields. Similarly, recovery of fertilizer P (RFP) was 18% for bush fields and 31% for infields over 3 years cropping. It is clear that degraded soils are poor in organic carbon, their response to fertilizer is less and the recovery of fertilizer applied is very low. Soil degradation was defined by FAO (2002) as the loss of soil productivity capacity in terms of decreased fertility, biodiversity and natural resources. Yield loss due to soil degradation in Africa has varied from 2% to 50% in the last 10 years (Scherr 1999). Bationo et al. (2006a) in Scherr (1999) and Oldeman et al. (1992) in a description of the level of degradation of arable soils in Africa and in the rest of the world have shown a proportion of 38% in the world and 65% in Africa. In the last 30 years, nutrients losses in Africa soils have been equivalent to 1,400 kg/ha N (urea), 375 kg/ha of SSP (phosphorus) and 896 kg/ha of KCl (potassium). In Niger, Henao and Baanante (2006) estimated nutrients losses of up to 56 kg/ha (NPK) during 2002–2004 cropping season.

Long term average of millet and sorghum grain yields were respectively, 400 and 190 kg/ha; but in 2002 and 2003 with respect to cereal production 3,336,956 and 3,561,660 tones were produced in Niger, and average millet and sorghum grain yields were, respectively, 461 and 476 kg/ha (FAO 2004). Research at ICRISAT (rapport annuel 1985) has shown that in the semi-arids zones of the Sahel where

Table 6.1 Long-term on-station trials at Sadore, Niger, West Africa

Type of trials	Site	Starting date	End
Long-term operational scale research	Sadore	1986	On-going
Long-term cropping system	Sadore	1993	On-going
Long-term crop residue management	Sadore	1982	On-going

annual rainfall is over 300 mm, nutrients are more limiting than water in crop production. At Sadore (Niger), with 560 mm of annual rainfall, 1.24 kg of millet grain per mm of water was harvested without fertilizer and 4.14 kg of millet grain per mm of water when fertilizer was used (Bationo et al. 2006a, b).

According to Oumou and Ed Heinemann (2006), Africa accounts for only 3% of the world fertilizer consumption with 13% of world arable soils and 12% of world population. Sub-saharan Africa (excluding South Africa) account for less than 1% in the world fertilizer consumption equivalent to 9 kg/ha compared Asia 148 kg/ha and Pacific. In 2002, fertilizer consumption in Niger was 1.1 kg/ha only; and one tone of fertilizer cost 400 $ compared to 90 $ in Europe thus making fertilizers less affordable.

The hill placement of small quantity (4 kg P/ha) can double millet grain yield. Micro-dose (4 kg P/ha) increased millet and sorghum grain yields up to 43–120% and farmer's income was improved through Warrantage by 52–134% in Niger.

Long-Term Management of Phosphorus, Nitrogen, Crop Residue, Soil Tillage and Crop Rotation in the Sahel

In 1986, a long-term soil fertility management trial was established by ICRISAT Sahelian Center to study the sustainability of pearl millet based cropping systems in relation to management of N, P, crop residue, rotation of cereal with cowpea and soil tillage. The data in Table 6.2 give the main treatments in this trial. This split-split-plot design, consisting of crop residue application or non-application of crop residue with half of the total crop residue produced in the plot. The sub-sub plot was with or without nitrogen application. Four replications were used in this experiment.

Results

Very low-measured against the environmental mean i.e. the mean yield performance of the different treatments-yields were observed on control plots for both grain and Total Dry Matter (TDM) yields. The other treatments showed high yields with better yield for T7 and T9 combining rotation, phosphorus with animal traction (AT) and hand cultivation (HC), respectively (Fig. 6.1).

Nitrogen effect was similar on both grain and TDM yields showing better yield on plot receiving N. But when annual rainfall was combined with yields, the higher TDM yield was observed with the higher annual rainfall; whereas millet grain yield was not subjected to higher rainfall. It shows that plant growth was better when

Table 6.2 Network collaborative trials indifferent sites in Niger (West Africa), 1999–2008

Type of trials	Site	Started	End
On-farm evaluation of cropping systems technologies	Sadore Karabedji Gaya	2003	Ongoing
Placement of phosphorus and manure	Karabedji	1999	Ongoing
Placement of phosphorus and PUE	Karabedji	2000	Ongoing
On-farm evaluation of cropping systems technologies, KKM project	Maradi	2008	Ongoing
On-farm evaluation of soil fertility restoration technologies	Karabedji Gaya	1999	Ongoing
Comparative effect of mineral fertilizers on degraded and non degraded soils	Karabedji	1999	Ongoing
Fertilizer equivalency and optimum combination of low quality organic and inorganic plant nutrients	Banizoumbou Karabedji Gaya	2001	Ongoing
Optimum combination of phosphate rock and inorganic plant nutrients	Banizoumbou, Gaya Karabedji, Sadore	2005	Ongoing
Corral experiment (demonstration)	Sadore	1988	Ongoing

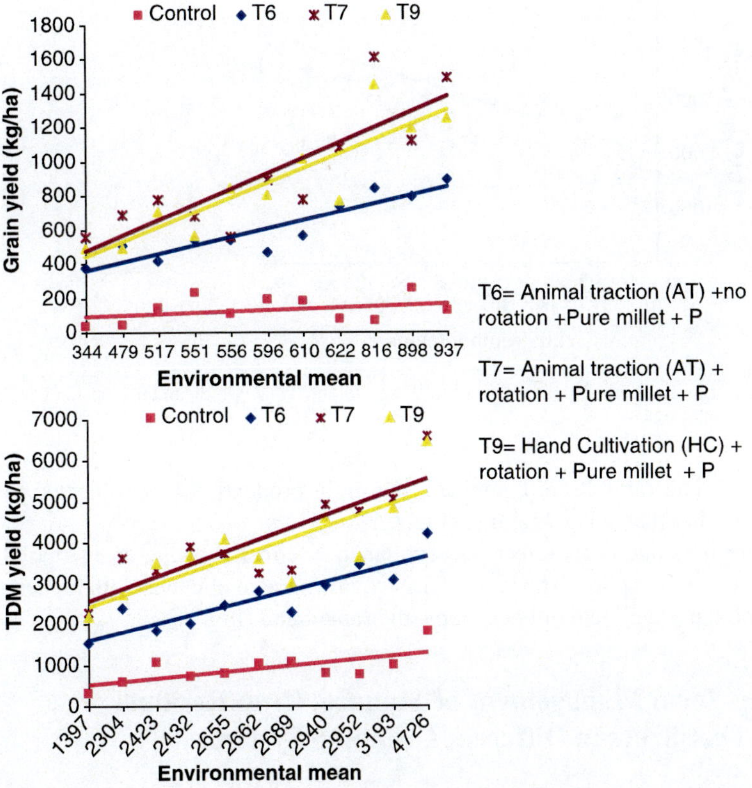

Fig. 6.1 Environmental mean on pearl millet grain and TDM yields under some technologies, Sadore, Niger, 1998–2008

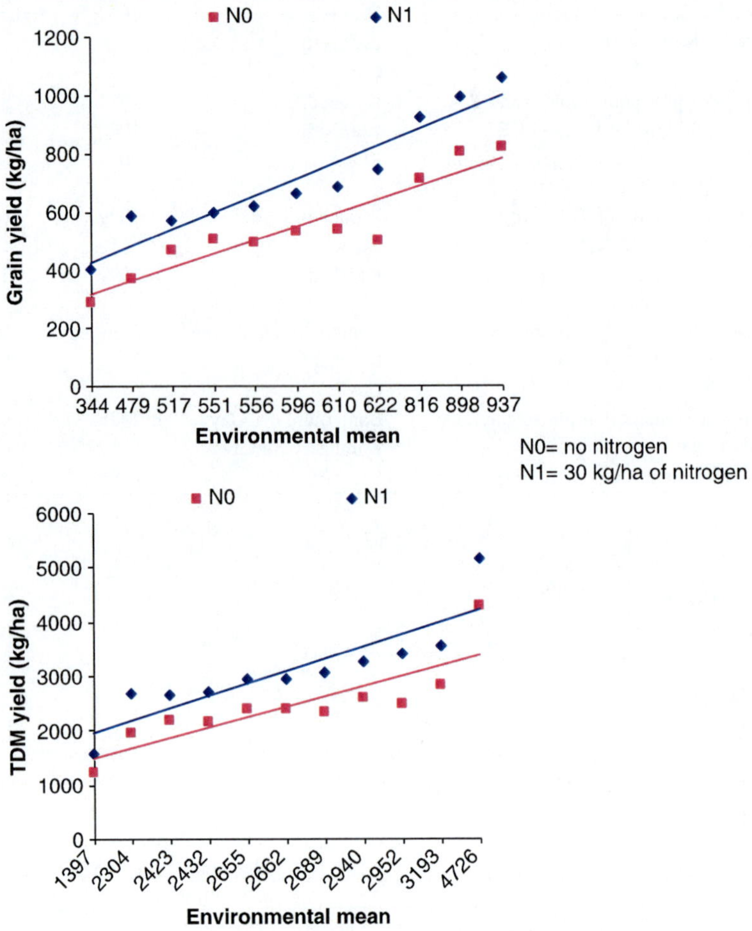

Fig. 6.2 Environmental mean on pearl millet grain and TDM yields under nitrogen use, Sadore, Niger, 1998–2008

nutrients and rainfall are combined; but grain production was subjected to other factors other than annual rainfall (Fig. 6.2).

Crop Residue (CR) effect was similar to N effect showing an increased yield when CR was applied. The higher TDM yield was also obtained with higher annual rainfall but grain yield did not follow the same trend (Fig. 6.3).

Long-Term Management of Manure, Crop Residues and Fertilizers in Different Cropping Systems

In 1993, a factorial experiment was initiated at the research station of ICRISAT Sahelian Center at Sadore, Niger. The first factor was three levels of fertilizers (0, 4.4 kg P + 15 kg N/ha, 13 kg P + 45 kg N/ha), the second factor was crop residue

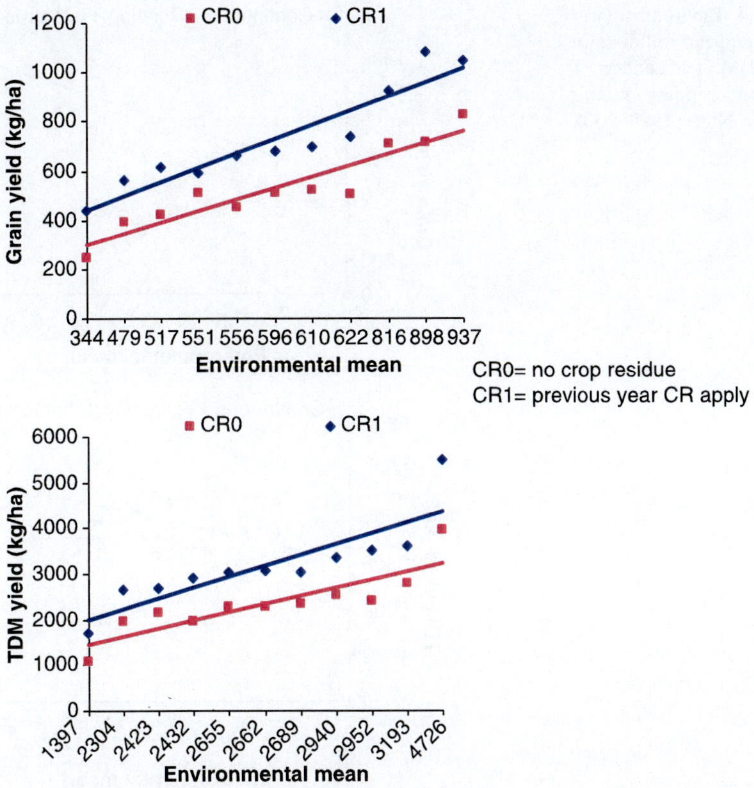

Fig. 6.3 Environmental mean on pearl millet grain and TDM yields under Crop Residue (CR) use, Sadore, Niger, 1998–2008

applied at (300, 900 and 2,700 kg/ha) and the third factor was manure applied at (300, 900 and 2,700 kg/ha). The cropping systems were continuous pearl millet, pearl millet in rotation with cowpea and pearl millet in association with cowpea. Three replications were used for this experiment.

Results

Among cropping systems rotation showed good performance compared to the two other systems for both grain and TDM yields during the whole period. There was no correlation between annual rainfall and millet grain yield and TDM production. The last 2 years gave low yields even though their annual rainfall was similar or higher than for other years (Fig. 6.4).

Although there was no significant difference between the different rates of CR, the higher yield was obtained with 2,700 kg/ha of CR. Annual rainfall was not correlated to millet grain and TDM yields (Fig. 6.5).

Fig. 6.4 Environmental mean on pearl millet grain and TDM yields under different cropping systems, Sadore, Niger, 1998–2008

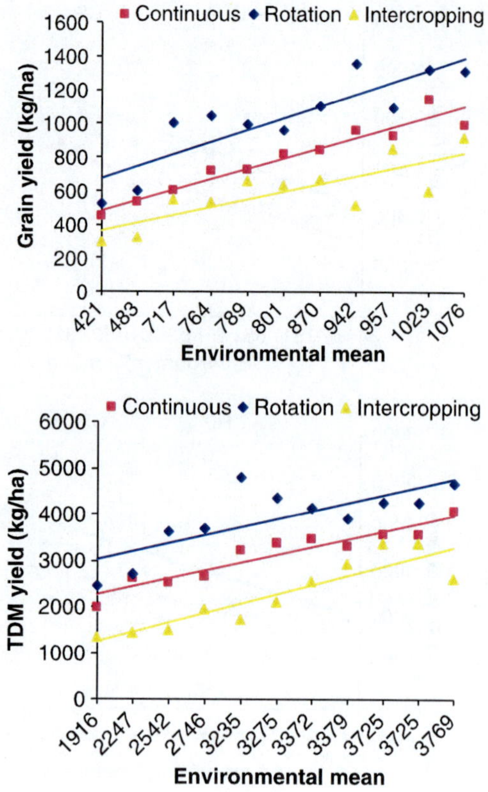

Manure application showed a large difference for millet grain and TDM yields with high yield for 2,700 kg/ha of manure. There was no correlation between annual rainfall and millet production with each factor varying independently (Fig. 6.6).

The use of fertilizer showed a very large difference between the control and the plot receiving P and N with very high yields obtained for high rate of application. It is clear that N and P application increased millet grain and TDM yield independent of annual rainfall (Fig. 6.7).

Long-Term Crop Residue (CR) Management

This crop residue (CR) trial established in 1982 showed a large cumulative effect on the soil (organic carbon, protection against erosion…) over the years. Four replications and four treatments consisted of traditional, sole application of CR, sole application of fertilizer (F) and CR+F. Each plot was split-plot and included rotation with half cowpea and half millet rotated every year. CR application involved leaving the previous year millet stover in the plot.

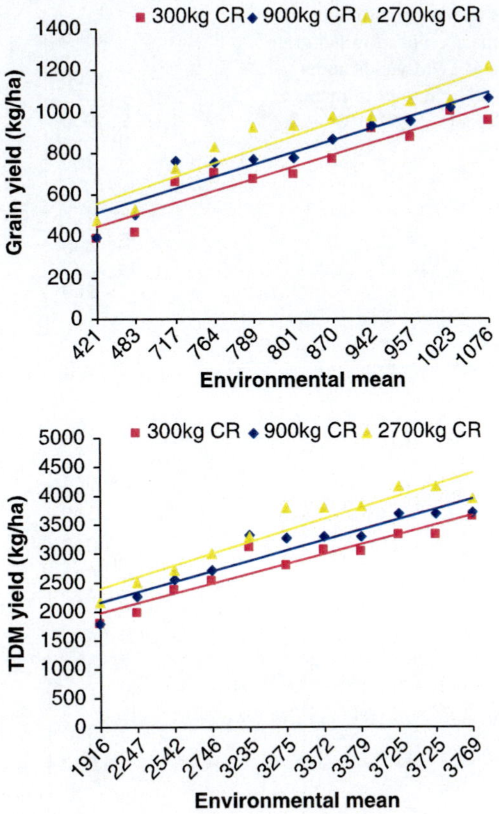

Fig. 6.5 Environmental mean on pearl millet grain and TDM yields under different rates of CR, Sadore, Niger, 1998–2008

Results

Data collected from 1984 to 2007 showed a significant difference between treatments. Yields in the control plot were low every year whereas Fertilizer + CR gave the highest yield with intermediary yields for the two other treatments (Figs. 6.8 and 6.9).

Better rainfall is supposed to give better yields but this was not always the case. Different situations were observed while combining yields and annual rainfall. With the lowest environmental mean of millet grain: 324 and 296 kg/ha, respectively, in 1993 and 2003, the corresponding rainfall was not very low: 541.7 and 534.3 mm, respectively, but enough for a good production as in 1996 where similar rainfall (543.9 mm) gave 1,362 kg/ha of grain yield. The lowest annual rainfall was recorded in 2000 with 392.7 mm of rainfall but millet grain yield was 753 kg/ha. In 1994 millet TDM yield was very low: 2,514 kg/ha with exceptionally high annual rainfall of 793.8 mm. It is clear that other internal factors such as rainfall repartition and drought spell affect more crop production than cumulative rainfall only. A survey on the daily rainfall demonstrated that when a contradiction occurred on the relation between annual rainfall and crop production, some explication could be found in its repartition.

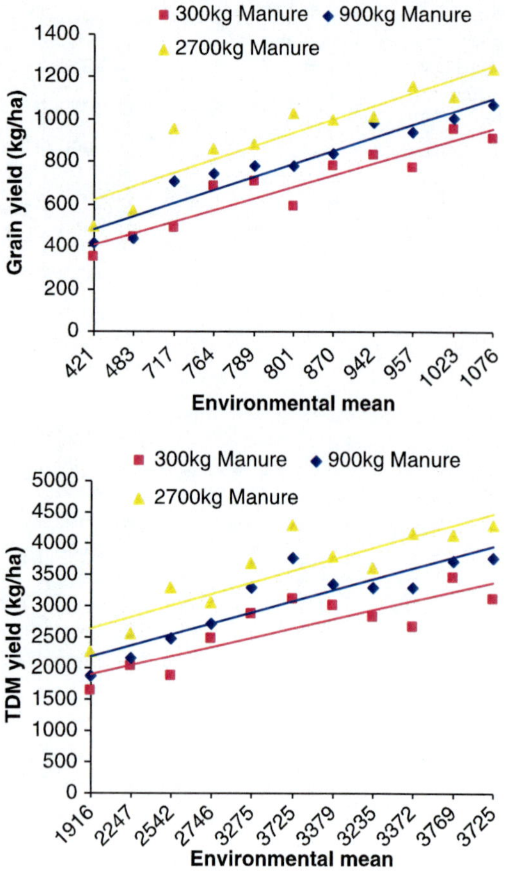

Fig. 6.6 Environmental mean on pearl millet grain and TDM yields under different rates of manure, Sadore, Niger, 1998–2008

For example the good yield in 1996 was related to a very good repartition although the annual rainfall was only 543.9 compared to 1994 where annual rainfall was 793.8. Similarly, the bad repartition of the same amount of annual rainfall gave very low yields in 1993 and 2003. The drought spells that occurred during the grain filling period affected grain yield more and those in the plant growth period affected more of TDM yield.

Other Trials in Different Sites

1. **Placement of phosphorus and manure (karabedji, Niger)**
   A complete factorial experiment was carried out with three levels of manure (0, 3, 6 t/ha), three level of P (0, 6.5 and 13 kg/P ha) using two methods of application (broadcast and hill placement).

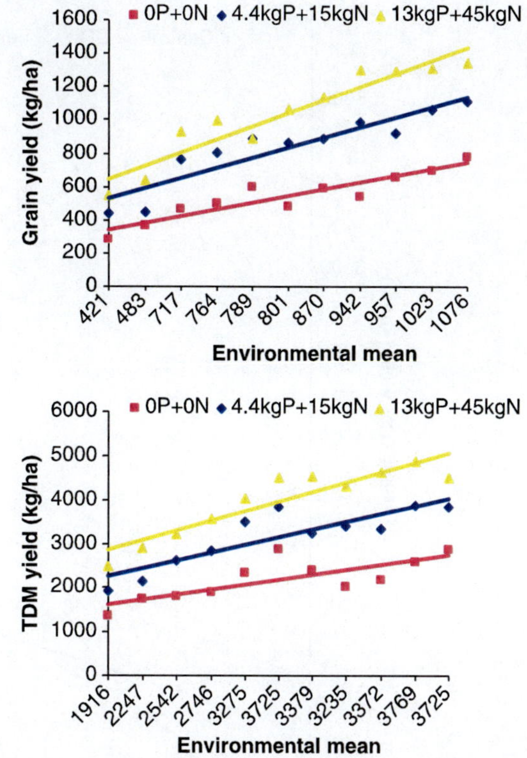

Fig. 6.7 Environmental mean on pearl millet grain and TDM yields under different rates of fertilizer, Sadore, Niger, 1998–2008

The data collected gave the response of millet to P and manure for the two methods of application. For pearl millet grain yield, the hill placement of manure performed better than broadcasting and with no application of P fertilizer.

The data for cowpea show also the same effect as for pearl millet. A very high yield of cowpea fodder can be produced with hill placement of manure and using it for feeding animals could allay affordability constraints.

2. **Placement of phosphorus and PUE (Karabedji, Niger)**
   Phosphate Rock (PR) was broadcast (bc) and/or hill placed (HP). For pearl millet grain P use efficiency for broadcasting SSP at 13 kg P/ha was low but hill placement of SSP at 4 kg P/ha gave a higher Phosphorus Use Efficiency (PUE). It also increased when NPK was hill placed at 4 kg P/ha as compared with PR broadcast. For cowpea fodder PUE was also better with P hill placed. The data clearly indicates that Phosphorus (P) placement can drastically increase P use efficiency and the placement of small quantities of water-soluble P fertilizers can also improve the effectiveness of phosphate rock.
3. **Farmer's evaluation of soil fertility restoration technologies (Karabedji and Gaya, Niger)**
   Past research results indicated a very attractive technology consisting of hill placement of small quantities of P fertilizers. With DAP containing 46% P_2O_5

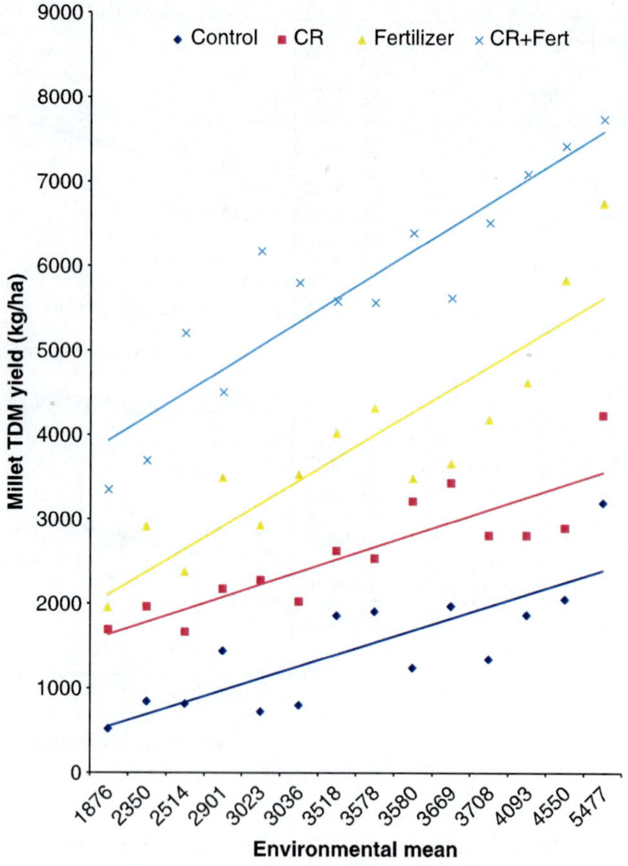

Fig. 6.8 Environmental mean on pearl millet TDM yield under CR and fertilizer uses, Sadore, Niger, 1984–2008

and a compound NPK fertilizer (15-15-15) containing only 15% P_2O_5, field trials were carried out by farmers on 46 plots per treatment at Karabedji to compare the economic advantage of the two sources of P for millet production. As hill placement can result in soil P mining another treatment was added consisting of application of phosphate rock at 13 kg P/ha plus hill placement of 4 kg P/ha as NPK compound fertilizer.

The data clearly showed that there was no difference between hill placement of DAP and 15-15-15 indicating that with the low cost per unit of P associated with DAP, this source of fertilizer should be recommended to farmers. The basal application of Tahoua Phosphate rock gave additional 200 kg/ha of pearl millet grain. The combination of hill placement of water-soluble P fertilizer with phosphate rock seems a very attractive option for the resource poor farmers in this region. The data showed the yield variation of each plot in farmers fields as

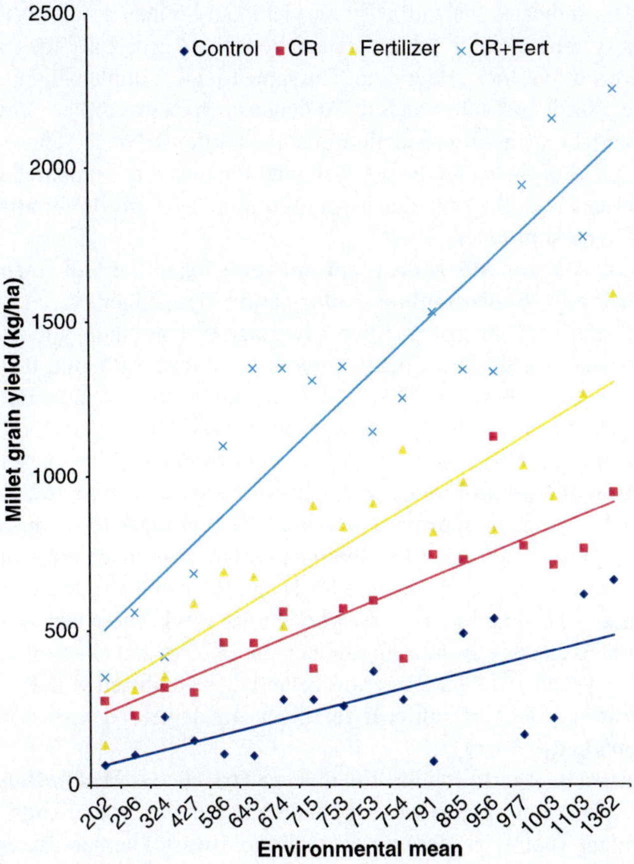

Fig. 6.9 Environmental mean on pearl millet grain yield under CR and fertilizer uses, Sadore, Niger, 1984–2008

compared to the farmer's practices and also clearly showed that the application of Tahoua PR with hill placement of water soluble P outperformed the other treatments in most instances.

At Gaya, with 50 plots per treatment the same effect was observed and DAP seemed better than NPK.

4. **Cropping systems and mineral fertilizers evaluation (Sadore, Karabedji, and Gaya, Niger)**

   Farmers' practices were compared to cropping systems consisting of pure millet crop with planting density at recommended level to be rotated the following year with a pure cowpea, and a fourth plot with millet and cowpea intercropped but not as farmer's practice. Here four lines of cowpea were rotated with two lines of millet. Tahoua Phosphate rock at 13 kg P/ha and NPK hill placed at 4 kg P/ha were applied for each plot except farmer's practice.

The data indicates that millet grain yield can be increased two to three fold with this system and higher biomass can be yielded especially for cowpea introduced to crop-livestock integration. The same trial was implemented at Karabedji and gave a similar effect as Sadore. At Gaya, cowpea was replaced by groundnut due to the lack of grain production in some particular years. These data shows how the yield of the technologies evaluated fluctuated as compared to the farmers' practices with the high density and rotation systems dominating the other systems in most instances.

5. **Combining PR and inorganic plant nutrients for millet and cowpea production (karabedji, Banizoumbou, Sadore and Gaya, Niger)**
The data indicated the comparative advantage of combining PR and inorganic plant nutrients for the low quality soils in the Sahel. Although the experiment started later than the others (2005), the sole application of inorganic P or PR P sources gave approximately the same yields in both millet grain and cowpea fodder while the combination of PR (50%) and inorganic P (50%) gave better yield.
6. **Interaction of N, P and manure (karabedji, Banizoumbou and Gaya Niger)**
A factorial experiment of manure (0, 2 and 4 t/ha), nitrogen (0, 30 and 60 kg N/ha) and phosphorus (0, 6.5 and 13 kg P/ha) was established in the three sites to assess the fertilizer equivalency of manure for N and P. The data showed a very significant effect of N, P and manure on pearl millet yield. The analysis conclusively demonstrated that P is the most limiting factor. Manure and nitrogen accounted for less in the system and their effect was better when combined with P.
7. **Comparative effect of mineral fertilizer on degraded and non-degraded soils (Karabedji, Niger)**
This is also a long-term soil fertility managed trial that started in 1999 and is still running. Mineral fertilizers were applied on two major type of soils: Farm close to the village (non-degraded) where household waste, human excreta and farm yard manure are commonly used and farm far from the village (degraded) and without any organic material added. The results showed a highly significant effect of the application of P on pearl millet grain and total dry matter yields. P significance was higher but nitrogen significance was linked to the availability of P or organic fertilizer in the soil. The system (degraded and non-degraded) showed a very high significance meaning that organic fertilizers are important. The data showed that the application of P increased significantly for both millet grain and total dry matter yields and it is more important in fertile soils where it is combined with organic fertilizer. Nitrogen alone was not significant but when combined with P the yield was significantly higher (Figs. 6.10 and 6.11).

Conclusion

The different soil fertility technologies tested in the different experiments and sites improved the millet grain and TDM yields. Results showed that the yield in the control plot is always lower than the environmental mean while improved

Fig. 6.10 Environmental mean of pearl millet grain on degraded and non degraded lands, Karabedji, Niger (1999–2006)

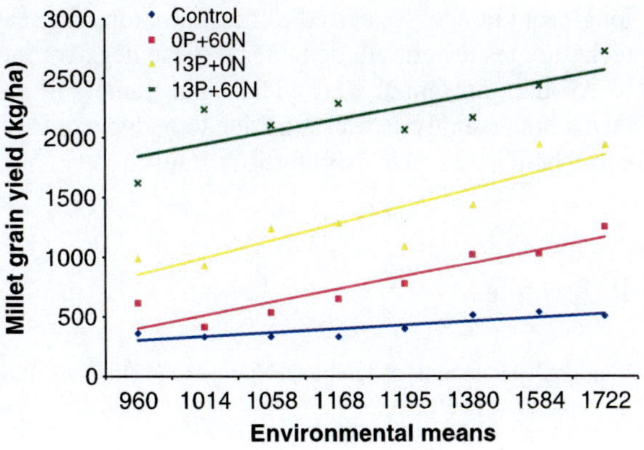

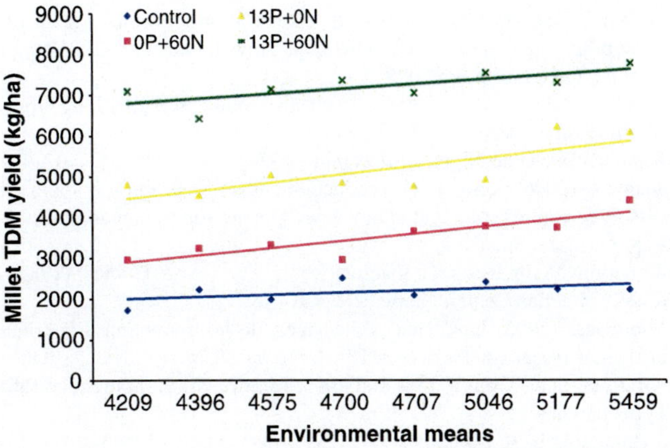

Fig. 6.11 Environmental mean of pearl millet TDM on degraded and non degraded lands, Karabedji, Niger (1999–2006)

technologies gave higher yields. Both organic and inorganic amendments increased millet production but their effects were different from one year to the other. Organic amendment performance was lower than inorganic but their combination was better.

Cropping systems experimented on showed very good performance for rotation where the previous leguminous crop gave additional nitrogen and organic amendment to the soil.

The annual rainfall and yields variations were not correlated meaning that factors other than rainfall affected millet production. For instance, in 2007 millet was planted on July 14th and rain stopped on September 14th. The lowest environmental mean was observed that year with only 2 months of rain while millet cycle was at least 3 months. In most of the cases where yields were low, rain stopped early or long drought spell occurred during the cropping season. The performance of the technologies tested will be better if these negative factors are addressed.

All data presented were yield data. Some soil samples were periodically collected but no analysis was done due to budget constraints. If soil analysis is done, it could help in assessing soil fertility status.

References

Bationo A, Hartemink A, Lungu O, Naimi M, Okoth P, Smailing E, Thiombiano L (2006a) Africa soils: their productivity and profitability of fertilizer use; background paper prepared for the African fertilizer summit, Abuja, Nigeria, 9–13 June 2006

Bationo A, Waswa B, Kihara J, Kimetu J (2006b) Nutrient cycling in agroecosystems: advantages in integrated soil fertility management in Sub-saharan Africa: challenges and opportunities, (Springer, edn.), Vol 76/2–3 Nov 2006

CILSS/AGRHYMET: Rapport final: Réunion de concertation technique sur les perspectives des productions et bilan céréalier ex-post 2004/2005 et prévisionnel 2005/2006 des pays du CILSS: Nouakchott du 7 au 10 novembre 2005

FAO (2002) Land degradation assessment in Drylands. Food and Agriculture Organization of the United Nations, Rome, 18 p

FAO (2004) Rapport spécial au Niger, 21 décembre 2004

Henao J, Baanante C (2006) Agricultural production and soil nutrient mining in Africa: implications for ressource conservation and policy development. International Fertilizer Development Center (IFDC), Muscle Shoals

ICRISAT (International Crop Research Institute for the Semi-Arid Tropics) (1985) Annual report 1984. ICRISAT Sahelian Center, Niamey, Niger

Moustapha Aboubacar (2003) Les céréales au Niger: de la production à la commercialisation; dans: forum1.inter-reseaux.net article 371

Nanga Jean, Grain de Sable.Com n°15 – mai 2005: Mali & Niger: la mondialisation néolibérale contre les plus pauvres

Oldeman LR, Hakkeling RTA, Sombroek WG (1992) World map of the status of human- induced soil degradation: An explanatory note, ISRIC, Wageningen, The Netherlands

Oumou Camara, Ed Heinemann (2006) Overview of the fertilizer situation in Africa; africa Fertilizer Summit, Abuja, Nigeria, 9–13 June 2006

Prudencio CY (1993) Ring management of soils and crops in West African semi-arid tropics: the case of the Mossi farming system in Burkina Faso. Agric Ecosyst Environ 47:237–264

Scherr SJ (1999) Past and present effects of soil degradation. In: Scherr SJ (ed) Soil degradation-a threat to developing-country food security by 2020, 2020 Discussion paper 27. International Food Policy Research Institute, Washington, DC, pp 13–30

Chapter 7
Effects of Organic and Inorganic Applications on Soil Bacterial and Fungal Microbial Communities Diversity and Impacts of Earthworms on Microbial Diversity in the Kabete Long-Term Trial, Kenya

M.M. Kamaa, H.N. Mburu, E Blanchart, L. Chibole, J.L. Chotte, C.N. Kibunja, and D. Lesueur

Abstract Soil fertility decline is increasingly leading to reduced food production worldwide. Over 70% of small holder farmers in the central highlands of Kenya are using crop manure, animal wastes and inorganic fertilizers to increase their farms' fertility and subsequent productivity. The dilemma with these practices is that less is known on the impact of these resources on the below ground biodiversity particularly the microbial communities which play a key role in determining soil quality. A study was carried out on a 32 year old long-term trial in Kabete, Kenya. These soils were treated with organic (maize stover at 10 t ha^{-}, farmyard manure at 10 t ha^{-}) and inorganic fertilizers (120 kg N, 52.8 kg P plus farmyard manure at 10 t ha^{-1} (N_2P_2 + FYM), 120 kg N, 52.8 kg P plus maize stover at 10 t ha^{-1} (N_2P_2 + R), 120 kg N, 52.8 kg P (N_2P_2), and a control (Nil and fallow) for over 30 years. We examined 16S rRNA gene and 28S rRNA gene fingerprints of bacterial and fungal communities, respectively, by PCR amplification and denaturing gradient gel electrophoresis (PCR-DGGE) separation. Bacterial community structure and diversity were negatively affected by N_2P_2, as evidenced by changes in the PCR-DGGE banding patterns. Bacterial community structure in the N_2P_2-treated soil was more closely

M.M. Kamaa (✉) • H.N. Mburu • L. Chibole • D. Lesueur
CIRAD, PERSYST Department, UPR "ETP", Tropical Soil Biology & Fertility, Institute of International Centre for Tropical Agriculture (CIAT-TSBF), ICIPE Duduville Campus off Kasarani Road, P.O. Box 823-00621, Nairobi, Kenya
e-mail: marykamaa2002@yahoo.com; Mary.Musyoki@uni-hohenheim.de

E. Blanchart • J.L. Chotte
Institut de Recherche pour le Développement IRD, Research Unit SeqBio, SupAgro Bâtiment 12, 2 Place Viala, 34060 Montpellier Cedex 1, France

C.N. Kibunja
Kenya Agricultural Research Institute (KARI), Kaptagat Road, Loresho, P.O. Box 57811–00200, Nairobi, Kenya

A. Bationo et al. (eds.), *Lessons learned from Long-term Soil Fertility Management Experiments in Africa*, DOI 10.1007/978-94-007-2938-4_7,

related to the bacterial structure in the untreated soil (fallow and Nil) than that in soils treated with a combination of inorganic and organic or inorganic fertilizers alone. For the fungal community the negative effect of N_2P_2 alone was not as adverse as for the bacterial community structure since the soils treated with N_2P_2 were closely related to those treated with N_2P_2 + FYM and N_2P_2 + maize stover. However, soils treated with organic inputs clustered away from soils amended with inorganic inputs. Organic inputs had a positive effect on both fungal and bacterial community structures with or without chemical fertilizers. Results from this study suggested that bacterial and fungal community structure was closely related to agro-ecosystem management practices conducted for over the past 30 years.

Keywords Organic fertilizers • Inorganic fertilizers • Bacterial communities • Fungal communities • Earthworms • DGGE • Long-term experiment

Introduction

In large parts of Kenya, food crop yields are very poor due to declining soil fertility as a result of continuous cropping with little or no replenishment of nutrients removed through either crop harvests or other losses such as leaching and soil erosion (Kibunja et al. 2010). With the liberalization of trade and introduction of Structural Adjustment Programs (SAP), unfavorable crop fertilizer prices and financial constraints have greatly contributed to the low level of fertilizer use in many small-holder' farms in the country (Waswa et al. 2007).

The Eastern and Central highlands small-holder farmers in particular, have been experiencing declining soil fertility and crop productivity. This has compelled researchers to place a high priority on increasing agricultural productivity and alleviation of poverty among small-holder farmers. Attainment of this goal calls for an Integrated Soil Fertility Management (ISFM) approach that is holistic and embraces the full range of driving factors of soil fertility right from the biological, physical, chemical and social-economic aspects of soil fertility (Kazombo-Phiri SFM 2005). In response to these problems, researchers in collaboration with farmers have set up trials on various soils management techniques for combating soil nutrient depletion. Technologies being actively and vigorously promoted to farmers in the region include: organic amendments, inorganic amendments or the combination of both organic and inorganic inputs. These technologies have been found to be both technically feasible and socially acceptable (Sanchez and Jama 2000; Jama et al. 1999). The emerging dilemma with this approach however, is that the impacts of these resources on soil microbial communities, which are key functional groups influencing agricultural production and ecosystem services, are not yet well documented.

Microorganisms play a key role in organic matter decomposition, nutrient cycling and other chemical transformations in soil, (Abbot and Murphy 2007). Microorganisms also immobilize significant amounts of carbon (C) and other nutrients within their cells. The total mass of living microorganisms (microbial biomass) therefore has a central role as a source, sink and regulator of the transformations of energy and nutrients in soil.

In both natural and agricultural systems, soil fauna can be important in organic matter decomposition, nutrient cycling, soil organic matter dynamics and pathways of water movement as a consequence of their feeding and burrowing activities. In the tropics, the effect of earthworms and termites, also referred to as "ecosystem engineers", are particularly pronounced because their activities influence the distribution of organic matter in the soil profile and the physical structure of soil (Lavelle et al. 2001).

Earthworms can be considered the most important "ecosystem engineers" in arable soil (Lavelle 1997), due to their lasting effects on soil physical and bio-chemical properties. Three broad functional groups of earthworms have been described by Bouché (1977) which represent particular adaptations developed to survive in litter (epigeics), soil (endogeics), or both (anecics). In the humid tropics, earthworms participating in soil processes are essentially endogeic, ingesting several times their own weight of soil daily. Endogeic earthworms live in the mineral soil and consume organic matter within the soil or at the soil-litter interface. Earthworm activity creates structures, casts and galleries that modify soil aggregation, porosity and the connections among pores. The digestion and further incorporation of non-digested soil organic matter (SOM) in casts have significant impacts on SOM dynamics at different scales of time, from hours to months and years (Lavelle et al. 2001). A major part of the beneficial effects of earthworm activity on soil properties is contributed to interactions with soil microorganisms. In nature earthworms play a major role in converting organic waste into simpler products and thereby improving the nutritive quality of the soil (Buckerfield and Kretzschmar 1992; Joy et al. 1992). During such a process, several soil microorganisms enter the earthworm gut. Many of them multiply well in the gut while a few are fully digested (Parthasarathi et al. 1997). All those micro-organisms that escape digestion in the gut proliferate well and return to the soil in the worm casts. However, these interactions are still poorly understood, including the effect of gut passage on the community structure of ingested soil microorganisms (Egert et al. 2004).

Land management practices have considerable impact on the size and dynamics of microbial populations. Intensification of agriculture has focused on the use of chemical and mechanical inputs, often at the expense of biologically-mediated processes. Both organic and inorganic sources of fertilizer have residue effects in the field. These effects are a vital component of sustainability because they smooth season-to-season variations in soil fertility and crop productivity, but they are difficult to asses quantitatively. Therefore, it is advantageous to undertake well-characterized medium-to-long term experiments rather than single-season trials, and to detail the interactions rather than averaging the responses over different seasons and environments (Tandon and Kanwar 1984). Long-term implies that primary objectives, treatment and management are not changed during the period under consideration, often regarded as at least 10 years (Laryea et al. 1995). An experiment was established in 1976 at Kabete and the primary objective of the trial was to identify appropriate methods for maintaining and improving the productivity of the soil through repeated use of inorganic fertilizers (in particular nitrogen and phosphorus), farmyard manure and crop residues under continuous cropping following the husbandry practices of small-scale farmers.

The challenge now remains to provide some understanding on the impact of these resources and earthworms on soil microbial communities. The work described here was carried out in the long-term field experiment at the National Agricultural Research Laboratories (NARL-KARI) Kabete, near Nairobi, Kenya, which provides a unique resource to investigate (1) the impact of long-term application of organic and inorganic resources on the soil bacterial and fungal communities and (2) the effect of earthworms introduction on the bacterial diversity of some of these soils.

Materials and Methods

The Study Site Description

The study was conducted as part of an ongoing project, at Kenya Agricultural Research Institute at the National Agricultural Research Laboratories (NARLs) station at Kabete, Kenya which is located at 36° 46′E and 01°15′S and at an altitude of 1,650 m above sea level (Kimetu 2002). The soils are mainly humic Nitisols (FAO 1990) that are deep and well weathered. The mean annual rainfall is about 950 mm received in two distinct annual rainy seasons; the long rains (LR) received between mid–March and June, and the short rains (SR) received between mid-October and December. The average monthly maximum and minimum temperature is 23.8°C and 12.6°C, respectively.

Choice of Treatments and Experimental Design

The experiment was established in 1976 as a randomized complete block design (RCBD) with 18 treatments replicated four times. For this study, only a selection of treatments in the original experiment receiving the organic and inorganic resources of interest was considered. The selection of these treatments (Table 7.1) was based on the crop yields data from the KARI yields data base.

Experimental Design for Bacterial and Fungal communities' Diversity

Soil samples were collected 6 weeks after planting following a zigzag systematic scheme, across the plots receiving treatments of interest (Table 7.1), by pushing a sterile soil auger at a depth of (0–10 cm). Six soil cores were taken and well mixed

Table 7.1 Selected treatments from the Kabete long-term experiment

Treatment	Fertilizer treatment	Organic amendments	Crop
1. Control	0 kg N, 0 kg P	0	Maize
2. Control + R	0 kg N, 0 kg P	Maize stover returned	Maize
3. FYM_2	0 kg N, 0 kg P	10 t/ha farmyard manure	Maize
4. N_2P_2	120 kg N, 52.8 kg P	0	Maize
5. N_2P_2 + R	120 kg N, 52.8 kg P	Maize stover returned	Maize
6. N_2P_2 + FYM	120 kg N, 52.8 kg P	10 t/ha farmyard manure	Maize
7. Fallow	Fallow	Fallow	Fallow

Key
R maize stover, *N* Nitrogen, *FYM* Farmyard manure, *P* Phosphorus

to make a composite sample of 20 g. The soils were then stored in zip lock plastic bags and transported in cooler boxes to the laboratory where they were stored at −80°C degrees for laboratory analysis (<3 weeks).

Experimental Design for Earthworms

Six soil cores were collected at a depth of 0–10 cm in from two treatments, Nil; no residue and Fertilizer at 120 kg N, 52.8 kg P plus Farmyard manure at 10 t ha^{-1} (N_2P_2 + FYM). Soil samples from each treatment were pooled together, air- dried and sieved (2 mm wire mesh). The soils were sub-sampled into six replicates per treatment of 300 g each and put in 24 boxes in a Complete Randomized Design (CRD) and maintained at 80% of its water holding capacity by adding 80 ml of deionized water. Fifteen adult endogeic geophagous earthworms were added in both treatments with a no earthworm control prepared in parallel. The boxes were incubated in the dark for 21 days, monitored and watered regularly in order to sustain earthworm activity after which the worms were removed and kept in alcohol, and the soils air-dried (without sieving). Twenty-five grams of soil was weighed and stored at −80°C for laboratory analysis.

DNA Extraction

Total genomic DNA was extracted from all soil samples by use of direct lysis extraction procedure (Martin-Laurent et al. 2001). A subsample of 0.25 g of soil was prepared using 0.5 g glass beads (106 μm diameter, Biospec Products) and 1 ml lysis buffer (100 mM Tris–HCl (pH 8.0), 100 mM NaCl, 1% (w/v) polyvinylpolypyrrolidone (PVPP), and 2% (w/v) sodium dodecyl sulfate (SDS)) in a 2 ml Eppendorf tube. The Eppendorff tubes were homogenized for 30 s at 1,600 rpm using a Biospec Mini-Bead- Beater cell disruptor, centrifuged at 14,000 g for 1 min at 4°C. The supernatant was collected, 1/10 volume of 5 M sodium acetate added and incubated on ice for 10 min, then centrifuged at 14,000 g for 5 min. The DNA was precipitated with one volume of ice cold isopropanol overnight. The DNA was washed with 70% ethanol and dissolved in double distilled water.

PCR Amplification of Bacterial 16S rRNA Gene

PCR amplifications were performed using the forward primer 338f with a GC clamp and reverse primer 518r (Øvreås et al. 1997). The total reaction mixture (25 μl) contained 2 μl of pure total DNA extract, 1X freeze dried bead (Ready-to-Go PCR beads, Pharmacia Biotech) containing 1.5 U of Taq polymerase, 10 mM Tris–HCl, (pH 9 at RT), 50 mM KCl, 1.5 mM $MgCl_2$, 200 μM of each dNTP and 1.0 μM of each primer and sterile distilled water. Two replicates were performed for each sample. A Primus 96 ^{plus} PCR system (MWG AG BIOTECH) thermal cycler was used for PCR amplification with 2 min at 94°C followed by 30 cycles of 30 s at 94°C, 30 s at 55°C and 1 min at 72°C. The first 20 cycles had an annealing temperature of 65°C, which decreased every cycle until a touchdown at 55°C. The primer extension was carried out at 72°C for 15 min. The PCR product (approximately 180 bp) was checked on a 1% agarose gel stained with ethidium bromide.

PCR Amplification of Fungal 28S rDNA Gene

PCR amplifications were performed using the forward primer 662f with a GC clamp and reverse primer 314r as described by Sigler et al. (2002). PCR amplifications were carried out in 25-μl reaction volumes with a Primus 96 ^{plus} PCR system (MWG AG BIOTECH). The reaction mixture was slightly modified by reducing the quantity of template DNA from 2.0 μl to 1.0 μl. Cycling conditions were as follows: initial denaturation at 95°C for 10 min followed by 49 cycles of denaturation at 95°C for 1 min, annealing at 50°C for 1 min, and extension at 72°C for 2 min followed by a final extension phase at 72°C for 10 min. The PCR product (approximately 348 bp) was checked for size on a 1% agarose gel stained with ethidium bromide.

DGGE Analysis

PCR products were separated using DGGE with a D-Code Universal Mutation system (Bio-Rad Laboratories). The PCR products were loaded on 8% (w/v) polyacrylamide (acrylamide-bisacrylamide [37.5:1]) gels with a denaturing gradient of 40–60% (100% denaturant contains 7 M urea and 40% formamide). The gels were run for 16 h at 75 V in Tris-acetate-EDTA buffer (1X) at 60°C with a pre-run of 10 min at 20 V. After migration, the DGGE profiles were stained with EtBr for 20 min, destained for 45 min and then scanned with Gel Doc ™ XR (BIO-RAD Laboratories). The gel images were captured using Quantity-One software version 4.6.0 (BIO-RAD Laboratories).

PCR-DGGE Gel Analysis

The DGGE patterns were transformed to graphs by the Bio-Rad Quantity One^{Tm} software: where each resolved band of the gel lanes was converted to a trait whose intensity is related to the amount of the corresponding DNA fragment. Total Lab 120 version 2006 software (Nonlinear Dynamics Ltd) was then used to calculate the percentage of similarity among lanes taking into account the migration distance and the relative intensity of all bands. The DGGE profiles were compared and unweighted-pair group method cluster analysis used to produce the dendograms.

Diversity Measurements and Data Analysis

The Shannon-Weaver indices (H') (Shannon and Weaver 1963) were calculated using the following equation:

$$H' = -\sum p_i \log p_i$$

$$\text{Where } pi = \text{n} / \sum \text{n}$$

Where:n equals the volume of a single band per lane∑n equals total volume of all bands per lane.The Shannon indexes were used to perform ANOVA analysis using GENSTAT Release 7.2 (Lawes Agricultural Trust 2007).

Results and Discussion

Bacterial and Fungal Communities as Affected by Organic and Inorganic Resources

Genetic finger printing by DGGE of eubacteria 16S rRNA amplified fragment showed few strong dominating bands and a number of faint unresolved bands (Fig. 7.1a), as compared to the PCR amplification of fungal 28S rRNA gene where most of the bands were distinct (strong) and well resolved (Fig. 7.1b). This could be

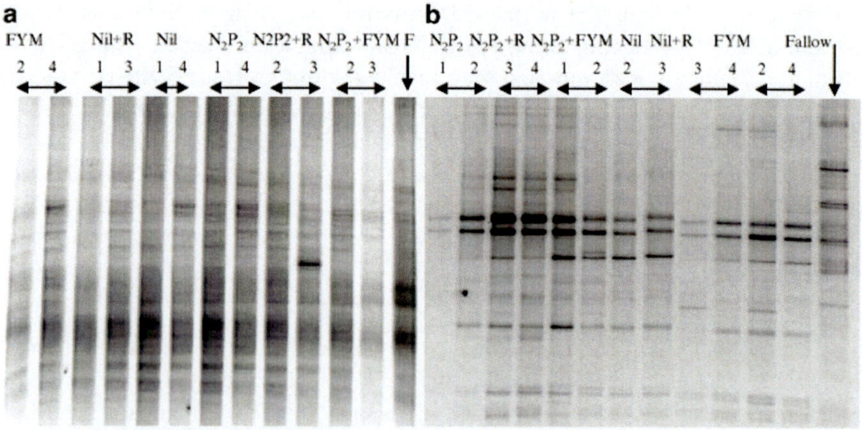

Fig. 7.1 DGGE profiles of 16S rRNA gene (**a**) and 28S rDNA gene (**b**) of soil sample from Kabete long-term experiment 2007. Numbers 1, 2, 3, and 4 represent selected treatment replicates. **Key** Nil–no input, Nil + R-Maize Stover, FYM_2-Farmyard manure at 10 t/ha, N_2P_2-chemical fertilizer at 120 kg N ha^{-1} and 52.8 kg P ha^{-1} N_2P_2 + FYM_2 – Chemical fertilizer at 120 kg N ha^{-1} and 52.8 kg P ha^{-1} with farmyard manure at 10 t ha^{-1}

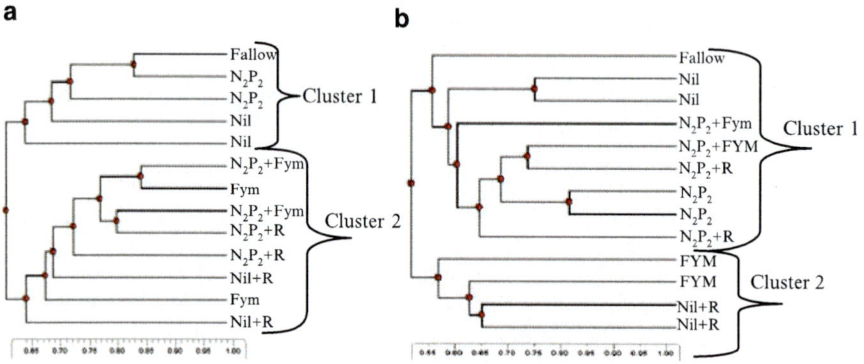

Fig. 7.2 Dendograms revealing relatedness of PCR-DGGE fingerprints from the gels in Fig. 7.1 (**a**) Bacterial, (**b**) Fungal clusters, using Dice's similarity coefficient

attributed to the higher heterogeneity of bacterial community than that of fungal communities. Cluster analysis of PCR-DGGE patterns differed in the fungal and bacterial community, although generally two major clusters were formed (Fig. 7.2). In the bacteria community, cluster analysis showed great similarity (65%) with two major clusters, those with organic inputs in one cluster and those with inorganic and no inputs in another cluster (Fig. 7.2a), while for the fungal communities, a moderately lower percentage (55%) with two major clusters was also observed, organically

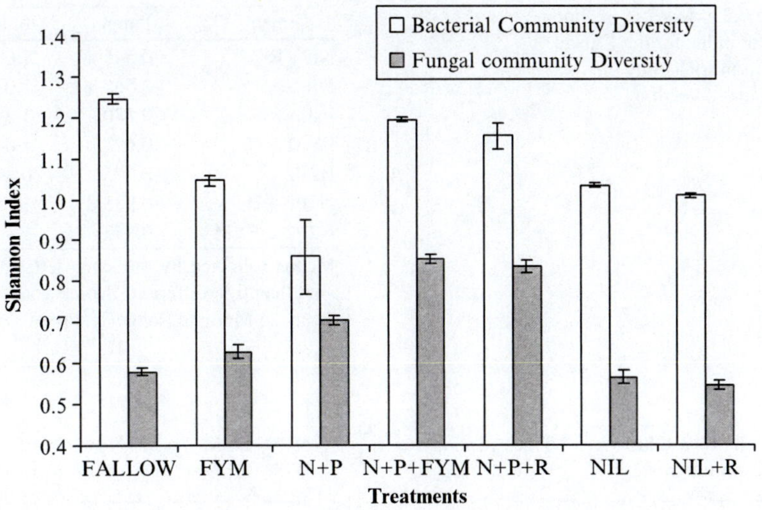

Fig. 7.3 Fungal and bacterial communities' diversity as affected by different treatments

treated soils clustered away from those with combined inorganic and organic inputs (Fig. 7.2b). Within the group of the combined inorganically and organically as well as that of inorganic treatments, the control and fallow treatments further grouped into a smaller cluster an indication that both organic and inorganic inputs had an influence on the structure of fungal and bacterial communities studied.

Shannon indices indicated a higher diversity of fungal and bacterial communities in the N_2P_2+FYM treatments. Lowest diversity in the bacterial community was in the N_2P_2 treated soils (Fig. 7.3) while for the fungal communities the lowest diversity was observed in the fallow and Nil treatments. This observation suggested a direct effect of organic and inorganic application on fungal and bacterial communities. The higher diversity in the N_2P_2 + FYM treated soils were correlated with enhanced biomass degradation thus high carbon contents from these soils compared to mineral fertilized soils (N_2P_2).

Statistical analysis of Shannon indices indicated that there were no significant differences ($p \leq 0.05$) between N_2P_2 + R, N_2P_2 + FYM and fallow treatments for bacterial communities (Table 7.2). A significant difference was however noted between these treatments and N_2P_2, Nil, Nil + R and FYM (Table 7.2). For the fungal communities no significant differences were found between N_2P_2 + R and N_2P_2 + FYM but significant differences were noted between fallow, Nil, Nil + R, FYM and N_2P_2. These results are in accordance with those obtained by Parham et al. 2002 on the evaluation of the culturable bacterial populations (Fig. 7.3).

Table 7.2 Means of Shannon Weaver index for the Kabete long-term trial study on microbial diversity

Treatment	Fungal	Bacterial
Nil + R	0.545 a	1.007 b
Nil	0.565 a	1.030 b
Fallow	0.580 a	1.245 d
FYM	0.627 b	1.045 bc
N2P2	0.705 c	0.860 a
N2P2 + R	0.835 d	1.152 cd
N2P2 + FYM	0.852 d	1.192 d

Means followed by the same letter are not significantly different from each other (Duncan Multiple Range Test $p < 0.05$)

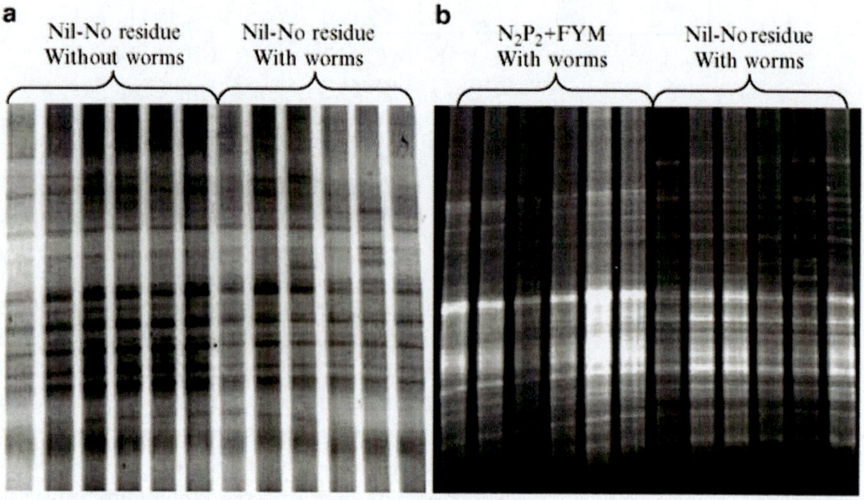

Fig. 7.4 DGGE profiles of bacterial 16S rRNA genes amplified from the soil treatments with and without earthworms

Bacterial Communities as Affected by Earthworms

Both soil treatments with and without earthworms produced considerable changes in bacterial community structure as illustrated in Fig. 7.4. Disappearance of some bands in the DGGE profiles of the soils inoculated with earthworms compared to DGGE profiles of soils not inoculated with earthworms indicated alteration of the abundance ratios of the bacterial groups harboring the DNA sequences revealed by these bands.

Further analysis performed with non-amended and amended soils confirmed that the bacterial community structure was modified by the addition of earthworms in both soils (Figs. 7.4 and 7.6a–d). However, in both soils the most important significance was

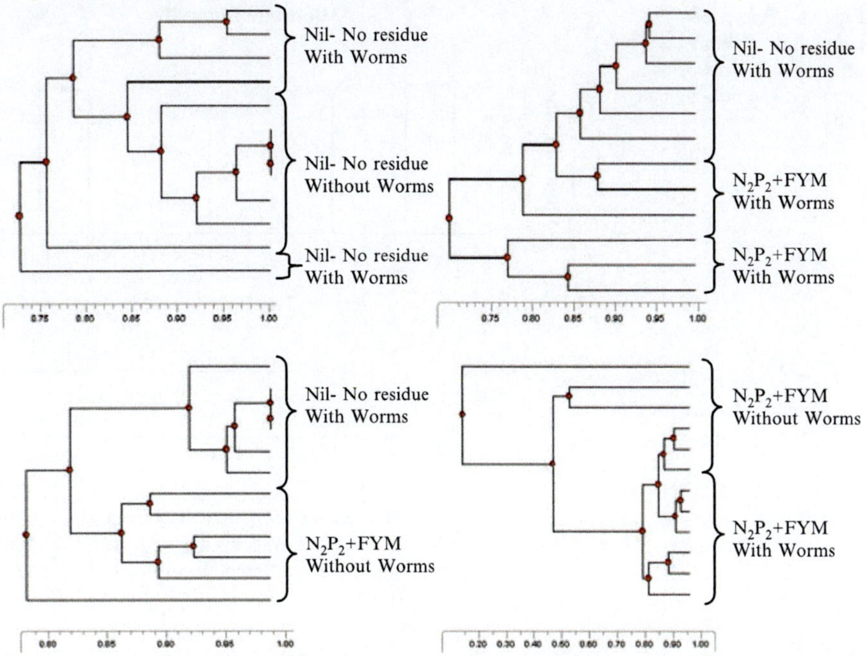

Fig. 7.5 Dendograms of DGGE profile similarities of the two soil treatments, Nil-No residue and N_2P_2+FYM with and without earthworms

produced by the introduction of earthworms in soil amended with N_2P_2+FYM (Figs. 7.5b, d and 7.6). Presence of earthworms had different effects on soil bacterial diversity. Earthworms in Nil- without residue soil treatment had no significant difference ($p < 0.05$) from the same soils without earthworms (Table 7.3). However introduction of earthworms in soils amended with N_2P_2 + FYM had a significant impact ($p < 0.05$) on the bacterial diversity as revealed by the anova analysis of the Shannon indices (Table 7.3, Fig. 7.6). Using ANOVA statistical analysis we found a significant influence of earthworms in soil amended with N_2P_2+FYM on the diversity of bacteria community. There was a decrease in bacteria communities when earthworms were introduced in amended soils which could be as a result of ingestion of the soil by the earthworms. This is in agreement with Wolter and Scheu (1999) who suggested that earthworm guts could be considered as favorable habitats for bacteria as there was an increased microbial numbers in the guts versus the soil, in which the earthworms were living. McLean (1997) also observed a decrease in the microbial community due to the invasion of *D. octaedra* in pine forest soil. Several studies have given hypotheses for differences between the microbial community structure of soil, earthworm gut, and casts are possible due to: (A) selective feeding of the earthworm on hot spots of microbial life (B) anoxic conditions in the gut lumen favoring microorganisms able to grow anaerobically (C) proliferation of microorganisms effectively exploiting the favorable physicochemical gut conditions;

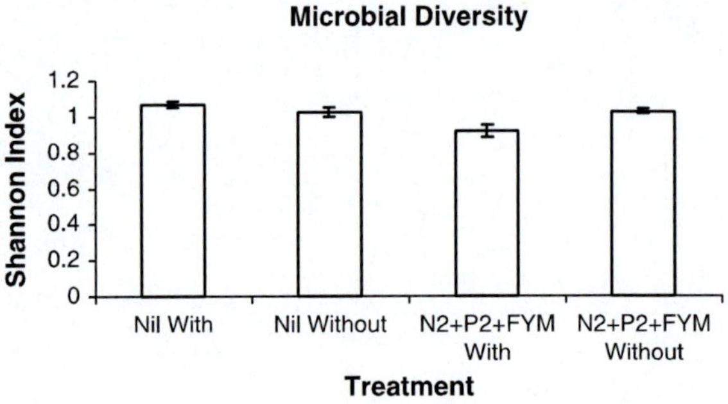

Fig. 7.6 Bacterial community diversity as affected by Earthworms in amended and non-amended soils

Table 7.3 Mean values of the Shannon indexes for the microbial diversity

	Earthworms	
Treatment	With	Without
N_2P_2 + FYM	0.92a	1.024b
Nil-no residue	1.028a	1.068a

*For each treatment, values (means of 6 replicates) followed by the same letter are not statistically different according to Fisher test ($p < 0.05$)

(D) differential lysis of microbes by digestive enzymes secreted by the earthworms; (E) inhibition of bacteria by inhibitory substances secreted by other bacteria (Kristufek et al. 1993; Brown 1995; Karsten et al. 1995; Wolter and Scheu 1999; Furlong et al. 2002 and Horn et al. 2003). Egert et al. (2004) observed that addition of beech litter to the soil markedly increased the differences between the investigated compartments for both bacteria and archaea, which is most likely due to preferential feeding of the earthworms on the supplemented litter and associated microorganisms. Ingested litter might also enhance some of the above mentioned processes with influence on the microbial community structure, because it means an additional input of nutrients into the gut possibly stimulating the secretion of digestive enzymes by the earthworm. For endogeic *Hormogaster elisae* earthworms, it was suggested that at least the amount of mucus added to the gut increases with increasing content of organic matter in the soil (Trigo et al. 1999). This means that addition of organic and inorganic materials (N_2P_2 + FYM) could have increased the effect of gut passage also by prolonging the time span the microbial community was exposed to the gut conditions.

Conclusion

Shifts in bacterial community structure following adaptation of soil management practices have been reported by other studies (Buckley et al. 2001; Buckley et al. 2003; Peacock et al. 2001, and Smit et al. 2001). Our results are in agreement with

Duineveld et al. (1998) who recognized a percentage similarity >18% in cluster analysis to indicate a significant change in community structure. We can therefore argue that long-term application of different organic and inorganic resources may be viewed as the main factor exerting a strong influence on the development of distinct microbial community structures. Nevertheless, plant type, soil type as well as climate are also important variables in determining the bacterial and fungal composition species.

Appearance of less numerous but strong bands in the fungal DGGE profiles would substantiate the hypothesis that a limited number of dominant and ecologically well-adopted fungal types were present in the long term experiment.

To-date, the combined use of chemical fertilizers and farmyard manure continues to consistently give higher yields than all the other treatments (Qureshi 1987; Swift et al. 1994; Kapkiyai et al. 1998; Kibunja et al., 2000; Kibunja 2007).

Our results are consistent with these findings where there was increased bacteria and fungal communities in soil treatments with N_2P_2 + FYM application, while Mariorana et al. 1992 observed a decrease in the average grain yields when N fertilization was simultaneously used over the long-term period, we also found the lowest bacterial communities in soil treatments with N_2P_2.

While the significance of the microbial community shifts following soil management practices remains to be recognized, it is evident that the combination of organic and inorganic amendments, in general, enriches the soil bacterial and fungal community and promotes diversity. We also found that the passage of soil bacteria in the mid-gut of the endogeic earthworms is largely soil or food derived, which could explain the significant differences in community composition in the two soil treatments (N_2P_2 + FYM and Nil-no residue) inoculated with earthworms. This could be explained by the fact that application of N_2P_2 + FYM resulted in an increase in the amount of organic matter and thus increase in the number of microbes which are responsible for organic matter breakdown (Parham et al. 2003; Ritz et al. 1997; Hopkins and Shiel 1996). Earthworms are like “Nature’s Tillers”. They would incorporate this organic matter into soil, ingest it, and excrete the nutrient rich casts on to and in soil. They would also increase the microbial biomass through multiplication in the earthworm guts which would influence a large number of important ecosystem processes, including nutrient acquisition, nitrogen and carbon cycling and soil formation.

The results provide information on important functional and structural soil microbial properties as influenced by organic and inorganic application at Kabete, Kenya long-term experiment. In conclusion, we showed that the bacterial and fungal community diversity and structure was impacted considerably by the fertilizer type. Combination of organic and inorganic resources increased soil bacterial and fungal diversity and resulted in a more even distribution than that in soil treated with inorganic fertilizers and in untreated control soils, these alterations were linked with the lower availability of biomass in inorganically treated soils.

Acknowledgement This work was supported by the "Microbes" project coordinated by Dr. J.L. Chotte and funded by l'Agence Nationale de Recherche (France). The authors are very grateful to The African Network in Africa (AfNet CIAT-TSBF) which is financially supporting the maintaining cost of this long-term trial and to Dr Andre Bationo and colleagues for their patience and relevant discussions about the trial.

References

Abbott LK, Murphy DV (2007) What is soil biological fertility? In: Abbott LK and Murphy DV (eds) Soil biological fertility-A key to sustainable land use in Agriculture. Spinger, Dordrecht, Netherlands, pp 1–15

Bouché MB (1977) Strategies lombriciennes. Ecol Bull 25:122–132

Brown GG (1995) How do earthworms affect microfloral and faunal community diversity. Plant Soil 170:209–231

Buckerfield JC, Kretzschmar A (1992) Earthworm population in dryland cropping soils under conservation tillage in South Australia. Soil Biol Biochem 24(12):1666–1672

Buckley DH, Schmidt TM (2001) Exploring the biodiversity of soil – a microbial rain forest. In: Staley JT, Reysenbach AL (eds) Biodiversity of microbial life: foundation of Earth's biosphere. Wiley-Liss, Inc., New York, pp 183–208

Buckley DH, Schmidt TM (2003) Diversity and dynamics of microbial communities in soils from agro-ecosystems. Environ Microbiol 5:441–452

Duineveld BM, Rosado AS, van Elsas JD, van Veen JA (1998) Analysis of the dynamic of bacterial communities in the rhizosphere of the chrysanthemum via denaturing gradient gel electrophoresis and substrate utilization patterns. Appl Environ Microbiol 64:4950–4957

Egert M, Marhan S, Wagner B, Scheu S, Friedrich WM (2004) Molecular profoling of 16S rRNA genes reveals diet-related differences of microbial communities in soil, gut, and casts of *Lumbricus terrestris* L. (Oligochaeta: Lumbricidae). FEMS Microbiol Ecol 48:187–197

FAO (1990) FAO-UNESCO Soil Map of the World Revised Legend. Soil resources report 60. Food Agriculture Organization: Rome, Italy

Furlong MA, Singleton DC, Whitman WB (2002) Molecular and culture based analyses of prokaryotic communities from an agricultural soil and the burrows and casts of the earthworm *Lumbricus rubellus*. Appl Environ Microbiol 68:1265–1279

Hopkins DW, Shiel RS (1996) Size and activity of soil microbial communities in long-term experimental grassland plots treated with manure and inorganic fertilizers. Biol Fertil Soils 22:66–70

Horn MA, Schramm A, Drake HL (2003) The earthworm gut: an ideal habitat for ingested N2O-producing microorganisms. Appl Environ Microbiol 69:1662–1669

Jama BA, Niang IA, Amadalo B, De wolf J, Rao MR, Buresh RJ (1999) The potential of improved fallows to improve and conserve the fertility of nutrient depleted soils of western Kenya In: Proceedings of the 6th Biennial KARI Scientific Conference (KARI), Nairobi, Kenya, pp 133–144

Joy EKD, Reddy MV, Reddy SM (1992) Mycoflora associated with the body surface, gut and casting of earthworm (Lampito mauritii Kinberg) in relation to soil in semi-arid tropical grassland. Environ Ecol 10:914–918

Kapkiyai JJ, Karanja NK, Woomer P, Quersh JN (1998) Soil organic carbon fractions in a long-term experiment and the potential for their use as a diagnostic assay in highland farming systems of central Kenya. Afr Crop Sci J 6:15–20

Karsren GR, Drake HL (1995) Comparative assessment of the aerobic and anaerobic microfloras of earthworm guts and forest soils. Appl Environ Microbiol 61:1039–1044

Kazombo-Phiri SFM (2005) Main causes of declining soil productivity in Malawi: effects on agriculture and management technologies employed to alleviate the problems. In: Omanya GO,

Pastern D (eds) Sustainable agriculture systems for the dry lands. Proceedings of the international symposium for sustainable dry land agriculture systems, Niamey, Niger, 2–5 Dec 2003, pp 59–66

Kibunja CN (2007) Impact of long-term application of organic and inorganic nutrient sources in a maize-bean rotation to soil nitrogen dynamics and soil microbial populations and activity. PhD thesis. University of Nairobi

Kibunja CN, Gikonyo EW, Wamaitha J, Thuranira EG, Nandwa S (2000) Sustainability of long term bean productivity using maize stover incorporation, residue fertilizers and manure. An Abstract of a paper presented during the 18th Conference and end of the Millennium Celebrations of the Soil Science Society of East Africa held on 4th–8th December, 2000, Mombasa Kenya. Oral paper 37

Kibunja CN, Mwaura FB, Mugendi DN (2010) Long-term land management effects on soil properties and microbial populations in a maize–bean rotation at Kabete, Kenya. Afr J Agric Res 5:108–113

Kimetu JM (2002) Nitrogen fertilizer equivalency values for organic materials of contrasting quality base on maize performance at Kabete, Kenya. Msc. Thesis, Kenyatta University, Kenya

Kristufek V, Ravasz K, Pizl V (1993) Actinomycete communities in earthworm guts and sorunding soil. Pedobiologia 37:379–384

Laryea KB, Anders MM, Pathak P (1995) Long-term experiments on alfisols and vertisols in the semi-arid tropics. In: Lal R, Stewart BA (eds) Soil management: experimental basis for sustainability and environmental quality. CRC Press, Boca Raton, pp 267–292

Lavelle P (1997) Faunal activities and soil processes: adaptive strategies that determine ecoisystem function. Adv Ecol Res 27:93–132

Lavelle P, Barois I, Brown GG, Blanchart E, Brussaard L, Decaëns T, Fragoso C, Jimenez J, Kajondo KK, Martinez M, Moreno A, Pashanasi B, Senapati B, Villenave C (2001) Earthworms as a resource in tropical agroecosystems. In: Subba Rao NS, Dommergues YR (eds) Microbial interactions in agriculture and forestry. Science Publishers, Enfield, pp 163–181. ISBN ISBN 1578081386

Lawes Agricultural Trust (2007) Genstat 7 release 1. VSN International, Wilkinson House, Jordan Hills Road, Oxford

Maiorana M, Convertini G, Di Bari V, Rizzo V (1992) Yield and quality of durum wheat (*Triticum durum Desf.*) under continuous cropping after 9 years of straw incorporation. Eur J Agronomy 1:11–19

Martin-Laurent F, Philippot L, Hallet S, Chaussod R, Germon JC, Soulas G, Catroux G (2001) DNA extraction from soils: old bias for new microbial diversity analysis methods. Appl Environ Microbiol 67:2354–2359

McLean MA (1997) Impacts of earthworm invasion on microbial activities and soil communities in a pine forest floor. PhD dissertation, University of Calgary

Øvreås L, Forney LJ, Daae FL, and Torsvik VL (1997) Distribution of acterioplankton in meromictic Lake Saelenvannet, as determined by denaturing gradient gel electrophoresis of PCR-amplified gene fragments coding for 16S rRNA. Appl Environ Microbiol 63:3367–3373

Parham JA, Deng SP, Raun WR, Johnson GV (2002) Long-term cattle manure application in soil. I. Effect on soil phosphorus levels, microbial biomass C, and dehydrogenase and phosphatase activities. Biol Fertil Soils 35:328–337

Parham JA, Deng SP, Da HN, Sun HY, Raun WR (2003) Long-term cattle manure application in soil. II. Effect on soil microbial populations and community structure. Biol Fertil Soils 38:209–215

Parthasarathi K, Ranganathan LS, Anandi V (1997) Predation of fungi by Lampito mauritii (Kinberg) reared on different substrates. Indian J Microbiol 37:41–42

Peacock AD, Mullen MD, Ringelberg DB, Tyler DD, Hedrick DB, Gale PM, White DC (2001) Soil microbial community responses to dairy manure or ammonium nitrate applications. Soil Biol Biochem 33:1011–1019

Qureshi JN (1987) The cumulative effects of N-P fertilizers, manure and crop residues on maize grain yields, leaf contents and some soil chemical properties at Kabete. pp. 12 in : National Maize Agronomy Workshop, Nairobi, 17–19th February 1987, KARI, Nairobi, Kenya

Ritz K, Wheatley RE, Griffiths BS (1997) Effects of animal manure application and crop plants upon size and activity of soil microbial biomass under organically grown spring barley. Biol Fertil Soils 24:372–377

Sanchez PA, Jama BA (2000) Soil fertility replenishment takes off in East and Southern Africa. ICRAF, Nairobi

Shannon C, Weaver W (1963) The mathematical theory of communication. University of Illinois Press, Urbana

Sigler V, William Ronald F, Turco (2002) The impact of chlorothalonil application on soil bacterial and fungal populations as assessed by denaturing gradient gel electrophoresis. Appl Soil Ecol 21:107–118

Smit E, Leeflang P, Gommans S, van den Broek J, van Mil S, Wernars K (2001) Diversity and seasonal fluctuations of the dominant members of the bacterial soil community in a wheat field as determined by cultivation and molecular methods. Appl Environ Microbiol 67:2284–2291

Swift MJ, Seward PD, Frost PGH, Qureshi JN, Muchena FN (1994) Long term experiments in Africa: developing a database for sustainable land use under global change. In: Leigh RA, Johnson AE (eds) Long term experiments in agriculture and ecological sciences, pp 229–251

Tandon HLS, Kanwar JS (1984) A review of fertilizer use research on Sorghum in India. International Crop Research Institute for the Semi-Arid Tropics. Research Bulleting No. 8. ICRISAT, Patancheru, 331 Andhara Pradesh, India

Trigo D, Barois MH, Gravin E, Huerta S, Irisson, Lavelle P (1999) Mutualism between earthworms and soil microflora. Pedobiologia 43:866–873

Waswa BS, Mugendi DN, Vanlauwe B, Kungu JB (2007) Changes in soil organic matter as influenced by organic residue management regimes in selected experiments in Kenya. In: Bationo A, Waswa B, Kihara J, Kimetu J (eds) Advances in integrated soil fertility management in Sub-Saharan Africa: Challenges and opportunities. Springer, Dordrecht, Netherlands, pp 457–469

Wolter C, Scheu S (1999) Changes in bacterial numbers and hyphal lengths during the gut passage through *Lumbricus terrestris* (Lumbricidae, Oligochaeta). Pedobiologia 43:891–900

Chapter 8
Cropping Systems in the Sudano-Sahelian Zone: Implications on Soil Fertility Management over Varied Seasons

A. Bationo, J. Kimetu, J. Kihara, Z. Traore, M. Bagayoko, V. Bado, M. Lompo, R. Tabo, and S. Koala

Abstract One of the approaches for achieving integrated soil fertility management (ISFM) to increase food production within the Sudano-sahelian zone (SSZ) is the testing (at farm level) the different cropping systems. The most common cropping systems in this zone involve growing several crops in association as mixtures or intercrops. A synopsis of the various cropping systems practiced in the SSZ and their implications on soil fertility is given in this chapter. Cropping systems discussed include: Intercropping, relay, sequential and crop rotation. Intercropping and

A. Bationo (✉)
Alliance for a Green Revolution in Africa (AGRA), Accra, Ghana
e-mail: A.Bationo@cgiar.org

J. Kimetu
Institute for Sustainable Energy, Environment and Economy (ISEEE),
Earth Sciences Building, University of Calgary, Calgary, AB, Canada

J. Kihara • S. Koala
Tropical Soil Biology & Fertility Institute of International Centre
for Tropical Agriculture (CIAT-TSBF), ICIPE Duduville Campus off Kasarani Road,
P.O. Box 823-00621, Nairobi, Kenya

Z. Traore
SRA, Cinzana, Mali

M. Bagayoko
CRRA Niono, Niono, BP 73, Mali

V. Bado
Institut de l'Environnement et de Recherche Agricolle (INERA), Programme
GRN/SP-Ouest Station de Recherche Agronomique de Farakô-Ba,
P.O. Box 910, Bobo-Dioulasso, Burkina Faso

M. Lompo
INERA, 03 BP 7192, Ouagadougou 03, Burkina Faso

R. Tabo
Forum for Agricultural Research in Africa (FARA), Accra, Ghana

A. Bationo et al. (eds.), *Lessons learned from Long-term Soil Fertility Management Experiments in Africa*, DOI 10.1007/978-94-007-2938-4_8,

rotation cropping systems have proved superior in several studies, therefore, there is need to have these technologies tested on-farm with close involvement of farmers to increase adoption rates. Farmer trials are indispensable in the scaling up of technologies and can be a good pillar in strengthening ISFM approaches. Consideration of the economic benefits of the technologies will be a vital component for the success of these technologies. Thus markets are increasingly becoming part of the research process since farmers have to trade off cash crops and the excess of their food crop produce. Participatory farmer research involving produce prices, post harvest processing, storage and farmer cooperatives should be conducted and may become an important incentive for successful adoption of the already proven technologies in the SSZ.

Keywords Cropping systems • Sudano-sahelian zone (SSZ) • Integrated Soil Fertility Management (ISFM) • Long-term • Farmer trials

Introduction

Despite the increasing global per capita food production, there has been an alarming decline in food production in the African continent over the past 25 years (Bationo et al. 2003). This has resulted to widespread malnutrition, a recurrent need for emergency food supply and an increasing dependence on food importation. According to Bationo et al. (2001a), about 28% of Africa's population is chronically hungry. This can be attributed to the relatively low growth rate for cereal grain yield (1%) as compared to the annual population growth rate which is about 3%. With large areas of forests, wetlands, river valley bottoms and grassland savanna being put under food crops, the food gap (requirements minus production) keeps widening. Agriculture-led development is fundamental to cutting off hunger, reducing poverty, generating economic growth, reducing burden of food imports and opening the way to an expansion of exports in the African continent.

As Bationo et al. (2001b) indicates, the sudano-Sahelian zone of West Africa (SSZWA) shelters the worlds' poorest people and 90% of the population live in the villages hence gain their livelihood from subsistence agriculture. However, over the last three decades, per capita food production has drastically reduced in this region. Soil fertility depletion has been described as one of the major biophysical root cause of declining per capita food production (Bationo et al. 2001a). This has been due to unsustainable production systems and continuous nutrient mining without sufficient external inputs for soil fertility replenishment.

Low soil fertility, which is a major constraint to increased food production in Africa, is a factor of natural and socio-economic aspects (Fig. 8.1) and is itself a large contributor to poverty and food insecurity. Low productivity of agriculture is related to the low quality of the soil resource base which on one hand has been due to inherent or induced deficiencies of major nutrients N, P, and K or low nutrient holding capacities, high acidity and low organic matter. Other constraints include low capacity for water retention, high P fixation and high erodibility and flooding (Rueler and Prins 1993). On the other hand low soil fertility is driven by socio-

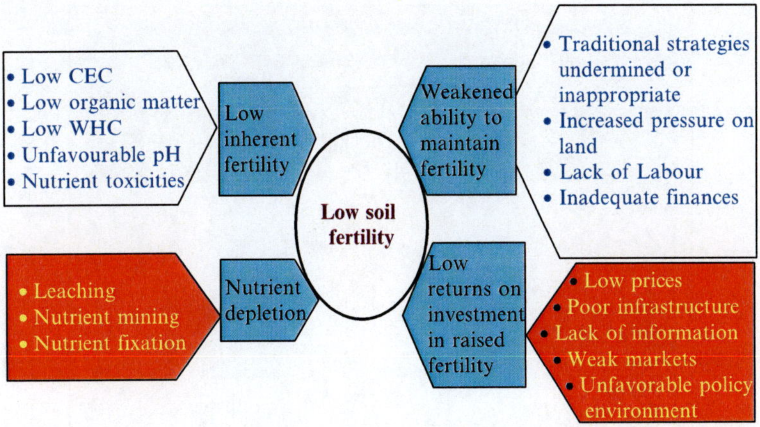

Fig. 8.1 Biophysico-chemical and socio-economic factors contributing to low soil fertility in Africa. Source: Bationo et al. 2004

economic factors, which include macro-economic policies, unfavorable exchange rates, poor producer prices, high inflation, poor infrastructure and lack of markets. These multiple causes of low soil fertility are strongly inter-related including the interaction between biophysical and socio-economic factors and hence need for a holistic approach in ameliorating the soil fertility constraints in sub-Saharan Africa (Murwira and Kirchmann 2003).

Approximately 65% (121 million hectares) of Africa's total land (187 million hectares) is degraded. Rates of nutrient depletion are particularly high in areas with favourable climates for crop production and high population densities. There is also high macronutrient loss as compared to application with many cropping systems reporting negative nutrient balances. For example, whereas 4.4 million tonnes per year are removed, only 0.8 million tonnes per year are applied, hence production takes place at the expense of the soil natural capital. Similarly, annual losses of K of upto 7% at farm level have been noted in semi-arid tropics of Africa.

Integrated soil fertility management (ISFM) has been regarded as a strategy that helps low resource endowed farmers mitigate many problems and the characteristics of poverty and food insecurity by improving the quantity and quality of food, income and resilience of soil productive capacity. The holistic approach to ISFM is shown in Fig. 8.2 and embraces the full range of driving factors and consequences-biological, physical, chemical, social, economic and political.

One of the approaches which has been sought by researchers to increase food production within the Sudano-sahelian zone is the testing of the different cropping systems. The most common cropping systems in this zone involve growing several crops in association as mixtures or intercrops. This practice provides the farmer with several options for returns from land and labor, often increases efficiency with which scarce resources are used, and reduces dependence upon a single crop that is susceptible to environmental and economic fluctuations.

In Sudanian zone, sorghum-based cropping systems are common. Pearl millet, maize, groundnut and cowpea are important components of this system. In the sahelian

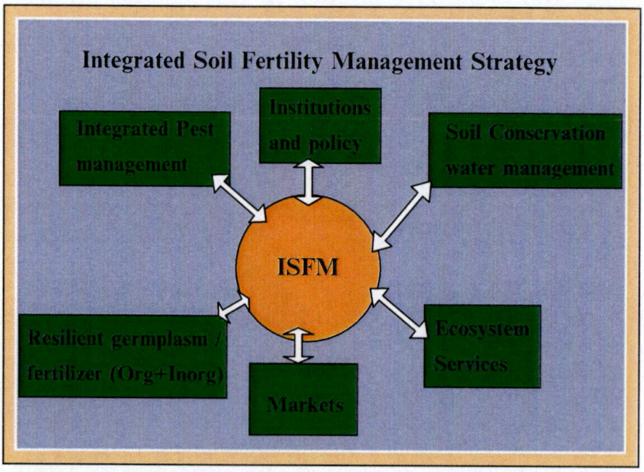

Fig. 8.2 Integrated Soil Fertility Management entry points with wider natural management concerns. Source: Bationo et al. 2004

zone, the cropping system is millet-based, with millet/cowpea and millet/groundnut being the important cropping patterns. While considerable information is available on fertilizer requirements for sole cropping of various crops, little is known on fertilizer requirement for intercropping.

Relay and sequential systems have been studied in recent years as an alternative to the traditional intercropping or sole cropping systems. Agronomic trials have been conducted to examine the performance of improved cultivars under relay and sequential systems and the potential of these alternative systems have been reported (Shetty 1984; ICRISAT 1984–1988; IER 1990).

Rotation of cereals with legumes has also been extensively studied in recent years. Use of rotational systems involving legumes for nitrogen fixation benefits is gaining importance throughout the region because of economic and sustainability considerations. The amount of N2 fixed by leguminous crops can be quite high though some workers have demonstrated that legumes can also deplete soil nitrogen (Rupela and Saxena 1987; Blumenthal et al. 1982; Tanaka et al. 1983). The quantity of N fixed in the sudano-sahelian zone concern the above ground part of the legume and, like in other parts of Africa, very little is known on the nitrogen fixed and retained within the roots. In the sudano-sahelian zone most of the legume biomass is used to feed livestock and not as green manure. This indicates that future research needs to focus more on the quantification of N fixed by the legume that is retained in root systems. Where much of the legume biomass is returned to the soil as green manure, a positive N balance is to be expected. However, this may be not true for grain legumes and fodders crops, where the bulk of above-ground legume material is removed from the systems. Nethertheless, many positive effects of grain legumes on succeeding crops are recorded (Bationo and Ntare 2000; Bagayoko et al. 2000). These suggest that factors other than the benefit due to biological nitrogen fixation need more research investigations.

In this chapter, emphasis is laid on intercropping, relay and sequential cropping, and rotation cropping systems.

Intercropping

Traditional intercropping systems cover over 75% of the cultivated area in the Semi Arid Tropics (Steiner 1984). As Norman (1974) observed, the principal reasons for farmers to intercrop are flexibility, profit, resource maximization, risk minimization, soil conservation and maintenance, weed control and nutritional advantages.

In the sudano-sahelian zone cereals such as millet and sorghum are traditionally intercropped with cowpea on small farms. In Niger, up to 87% of the millet area is intercropped (Swinton et al. 1984). A study by Norman (1974) in northern Nigeria has shown that only 8% of the area was planted to sole sorghum, while about 50% of the area was planted with sorghum in intercrop. Similar findings are reported by Shetty et al. (1987) for Mali and by Fussel and Serafini (1985) for Nigeria, Niger, Burkina Faso and Mali. The different authors have clearly underlined the importance of intercropping in view of its maximum utilization of resources and stabilization of yields. The most common associations are cereal/cowpea, cereal/groundnut, and cereal/cereal such as millet/sorghum/maize and millet/sorghum/cowpea. In these systems pearl millet is normally sown first and acts as the dominant crop.

In the cowpea/cereal intercropping, the cowpea and cereal are usually planted in alternating rows, but research at the International Institute of Tropical Agriculture (IITA) has shown that planting four rows of cowpea to two rows of cereal is more productive. The cereal is planted first, followed by the cowpea. The fast growth and spreading habit of traditional cowpea varieties suppress weeds, and soil nitrogen is increased which improves cereal growth. The two crops are harvested at different times, spreading available labor.

Norman (1974) concludes that mixed cropping is a strategy for the farmers' profit maximization and risk minimization. Production and income stability are important features of the systems (Abalu 1976) which also alleviate seasonal labor peaks. The most advantageous intercropping systems exploit the temporal differences between crops through a temporal separation of the most competitive growth periods (Fussell 1985; Serafini 1985; Fussell et al. 1986; Baker 1979). In the millet/cowpea intercropping systems, cowpea planted 2, 3 or 4 weeks after the millet will exploit the moisture and nutrients used by the millet without always a significant reduction in the cereal yield because the legume had a longer growth duration (Fig. 8.3). The system is flexible and in a year of early drought the staple cereal will suppress the non-cereal crop. On the other hand when cereal established poorly, the density of the second crop can be increased to cover up for the available space (Fussell et al. 1986).

The yield advantages of intercrop systems vary from 10% to 100% in millet (Fussell and Serafini 1985). The data in Table 8.1 indicate that in western Niger, the combination of cowpea with millet has resulted in production advantages of 10–40% over 4 years, and in Mali by 100% in maize/millet intercrop. Yield stability has been proposed as a major advantage of intercropping (Willey 1979; Willey et al. 1985;

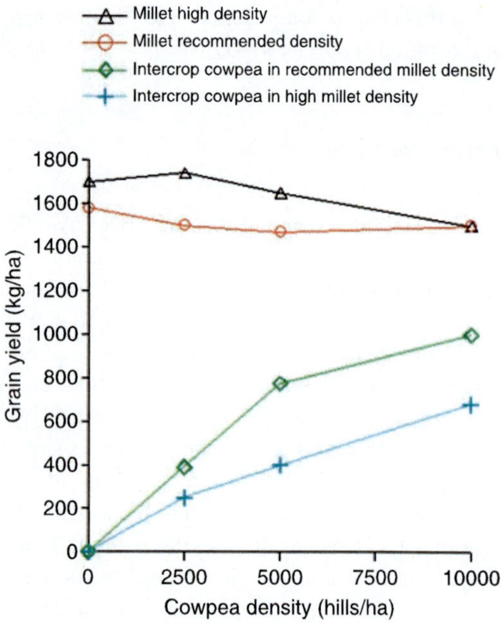

Fig. 8.3 Pearl millet grain yield and cowpea hay response in a pearl millet/cowpea intercropping system at three cowpea densities, ISC, Niger, 1985

Steiner 1984) as farmers want to rely on management practices that increase yields, when this is possible, while improving the stability of the production in both good and poor rainfall years. Stability may result from one crop compensating for the poor performance of the other crop. Baker (1980) has compared the relative stability of intercropping and sole cropping using the stability analysis of Finlay and Wilkinson (1963) and found that in the groundnut/cereals systems in northern Nigeria, intercropping systems were found to be more stable.

For millet/groundnut systems a yield advantage between 28% and 53% is reported in Table 8.2. Although traditional intercropping systems cover over 75% of the cultivated area in the sudano-sahelian zone, there is a scarcity of information on the efficiency of fertilizers under these systems. The number of days before planting the second crop will depend on the importance of the next rain after the first cereal crop has been planted. With a basal application of P fertilizers the cereal crop growth is rapid and can suppress completely the second crop if its planting occur after 3 weeks of sowing the cereal crop. In contrast, if the legume crop is planted early, it will compete more with the cereal crop for light, water and nutrients and can significantly reduce the yield of the cereal crop. The data in Table 8.3 indicate that there is differential responses among cowpea cultivars to soil fertility and research at International Center for Research in the Semi-Arid Tropics (ICRISAT) Sahelian center clearly indicated that cereal yields respond better to fertilizer when cowpea density is low (ICRISAT 1984–1988). Thus cowpea cultivars best suited to intercropping with millet should not strongly compete with the cereal crops but should themselves maintain high yields of both grain and fodder in an intercrop situation. The data from the intercropping of groundnut with pearl millet indicate that in some situations yield disadvantage of intercropping can be as high as 18% (Table 8.4).

Table 8.1 Intercropping effect from trial means of factorial experiments which included other management factors such as intercrop density, fertility, planting data and harvest time of intercrop, Niger and Mali

Year Type of experiment	Location	Rainfall (mm)	Cowpea ha/maize grain (kg/ha)	Millet grain (kg/ha)	Land equivalent ratios		
					Cowpea	Millet	Total
Millet/cowpea intercrop							
1982 Density x variety x intercrop proportion	Sadoré, Niger	372	318	277	0.58	0.69	1.27
1983 Intercrop proportion x rotations	Sadoré, Niger	599	768	385	0.48	0.69	1.17
1984 Intercrop proportions and rotations	Sadoré, Niger	216	22	435	0.09	1.16	1.25
1985 Intercrop proportion x rotations	Sadoré, Niger	495	734	648	0.48	0.72	1.20
1985 Variety x intercrop	Sadoré, Niger	495	360 (261)**	921	0.31 (0.50)	0.93	1.24(1.43)
1979 Density x intercrop	Multilocation, Mali	NA	158	854	0.96	0.21	1.17
1980 Density x intercrop	Multi-location, Mali	NA	1,460	980	1.04	0.50	1.54
1981 Intercrop harvesting schedule	Multi-location, Mali	NA	4,050	800	1.12	0.83	1.95
Maize/millet intercrop:							
1982 Intercropping study	Cinzana, Mali	579	2,010	840	1.12	0.57	1.69
1982 Intercropping study	Sikasso, Mali	1,046	1,800	880	1.04	1.06	2.10
1983 Planting date x intercropping	Sikasso, Mali	756	590	1,920	0.29	0.82	1.11

*Hay yields are fresh weights
*Cowpea grain yields

Table 8.2 Yields of pearl millet and groundnut in sole crops and intercrops, and resultant land-equivalent ratio (LER) at Tara, Niger, rainy season 1989

Treatment[a]	Groundnut pods ($t\ ha^{-1}$)	Millet grain ($t\ ha^{-1}$)	LER[b]	Groundnut haulms ($t\ ha^{-1}$)	Millet straw ($t\ ha^{-1}$)	LER
Sole crop						
Groundnut (28–206)	1.29			2.62		
Groundnut (47–16)	0.99			3.13		
Groundnut	1.40			2.59		
[ICGS(E)11]		1.29			3.70	
Millet (CIVT)						
Intercrop						
CIVT and 28–206	0.71	1.20	1.48	1.28	2.95	1.29
CIVT and 47–16	0.66	1.04	1.46	1.44	3.05	1.28
CIVT and ICGS(E)11	0.71	1.31	1.53	1.22	3.32	1.37
SE	±0.08	±0.05		±0.17	±0.17	
CV(%)	16.6	16.2		16.8	19.0	

[a]Randomized complete block design with 4 replications. Millet planted at 1 × 1 m and groundnut at 50 × 100 cm
[b]LER = Sum of ratios of yield of each crop in mixture over yield of sole crop

Table 8.3 Effect of fertilizer on the yield of five cowpea cultivars intercropped with millet at Gobery, Niger, rainy season 1987

	Yield ($t\ ha^{-1}$)							
	Cowpea hay				Millet grain			
Treatment	F1[b]	F2	F3	Mean	F1	F2	F3	Mean
VITA 3	0.12	0.07	0.13	0.11	0.16	1.19	1.00	0.78
TVX4659-03E	0.11	0.15	0.17	0.14	0.44	1.14	1.28	0.95
TN88-63	0.21	0.26	0.26	0.24	0.32	1.06	1.60	0.99
TN3-78	0.13	0.10	0.05	0.09	0.30	1.03	1.01	0.91
Sadore Local	0.64	0.47	0.46	0.52	0.14	0.88	1.52	0.84
Sole millet	–	–	–	–	0.29	0.96	1.36	0.78
SE	±0.03		±0.02		±0.15		±0.06	
Mean	0.18	0.15	0.16		0.33	1.04	1.35	
SE	±0.01			±0.06				
CV (%)	41			33				

[a]Split plot with 4 replications; plot size 36 m^2
[b]F1 = no fertilizer added; F2 = 20 kg $P_2$05, 22.5 kg N; F3 = 40 kg $P_2$05 and 45 kg N ha^{-1}

Relay and Sequential Cropping Systems

In the Sudanian zone with a longer growing season and higher rainfall there is greater opportunity than in the Sahelian zone to manipulate the systems with appropriate genotypes and management systems. Field agronomic trials have been conducted to examine the performance of the cultivars under relay and sequential

Table 8.4 Grain yields (t ha^{-1}) of pearl millet and groundnut sole-cropped or intercropped at Sadoré, Niger, rainy season 1987, 1988

Crop/ cultivar	Sole pearl millet		Intercropped groundnut		Sole groundnut		LER[b]	
	1987	1988	1987	1988	1987	1988	1987	1988
Pearl millet (cv CIVT)	0.93	1.71						
Groundnut								
55–437	0.44	0.90	0.30	0.61	0.34	1.14	1.35	1.07
ICGS(E) 21	0.47	0.93	0.26	0.93	0.41	1.09	1.13	1.39
ICGS(E) 13	0.37	0.86	0.23	0.84	0.43	1.31	0.93	1.14
ICGS(E) 20	0.38	0.92	0.21	0.47	0.45	1.16	0.88	0.95
ICGS(E) 19	0.43	1.13	0.15	0.67	0.42	1.23	0.82	1.20
SE	±0.09	±0.18	±0.04	±0.12	±0.07	±0.24		
CV (%)	38	33	36	35	34	41		

[a]Randomized-block design with 4 replications; plot size 36 m^2
[b]LER – combined land equivalent ratio, calculated using trial means

systems and reveal the potential of these alternative systems over traditional sole or mixed cropping (Shetty 1984; ICRISAT 1984–1988). In Mali, by introducing short season sorghum cultivars in relay cropping with other short duration cowpea and groundnut cultivars, substantial yields of legumes and sorghum were obtained as compared to traditional systems (IER 1990).

In the sahelian zone Sivakumar (1988) analyzed the date of the onset and ending of the rains, and the length of the growing period. He found that an early onset of the rains offers the probability of a longer growing season while delayed onset results in a considerable shorter growing season. The above analysis suggests that even for the sahelian zone, cropping management factors using relay cropping can increase soil productivity. The data reported by Sivakumar et al. (1990) indicate that the advantage of the millet-cowpea system under early rains treatments is reflected in the water use and the water use efficiency value. In 1986 these crops together used 597 mm as opposed to 374 mm for the sole crop of millet and in 1987 total water used by millet and cowpea in relay cropping was 585 mm, in contrast to 398 mm by the sole crop millet. Sivakumar et al. (1990) concluded that the potential of the rainy season in years with early onset of rains in the sahelian zone can be exploited by establishing a second crop of cowpea after the millet. In 1987 as an example a normal cropharvest of 810 kg grain and 3,780 kg straw was obtained but in addition 480 grain/ha of cowpea and 380 kg hay/ha was harvested (Table 8.5).

Crop Rotation

In the mixed cropping systems legumes are grown between cereals at very low densities resulting in poor legume yields (Ntare 1989; Reddy et al. 1992). For example, cowpea grain yields are very low, varying between 50 and 300 kg ha^{-1} in marked

Table 8.5 Grain yield (kg ha^{-1}), water use (mm) and water use efficiency (WUE) (kg ha^{-1} mm^{-1}) of cropping systems during the 1986 and 1987 rainy seasons

	Millet					Cowpea				
				WUE					WUE	
Treatment	Grain	Straw	Water use	Grain	Straw	Seed	Hay	Water use	Seed	Hay
1986 rainy season										
Early rains (irrigated)	1,730	4,100	398	4.4	10.3	50	680	237	0.2	2.9
Early rains (rainfed)	1,210	3,650	374	3.2	9.8	–	420	223	–	1.9
SE (%)	100	410	7	0.31	1.2	–	10	4		0.05
CV (%)	10	15	3	12	17	50	3	3		3
1987 rainy season										
Early rains (irrigated)	1,050	5,370	398	2.7	13.7	480	380	187	2.6	2.0
Late rains	810	3,780	316	2.6	12.1	–	–	–	–	–
SE (%)	90	420	16	0.29	1.69	68	67	5	0.33	0.32
CV (%)	19	18	9	22	26	29	35	6	26	30

Table 8.6 Millet grain and total dry matter yield at harvest as influenced by millet/cowpea cropping system at Sadore (Niger). Source: Bationo and Ntare 2000

	Grain yield			Total dry matter yield		
	1996	1997	1998	1996	1997	1998
Cropping system	kg ha^{-1}					
Continuous millet	937	321	1,557	4,227	2,219	6,992
Millet after cowpea	1,255	340	1,904	5,785	2,832	8,613
P > F	<0.001	0.344	<0.001	<0.001	<0.001	<0.001

contrast to yield of over 1,000 kg ha^{-1} obtainable on research stations and by large-scale commercial enterprise in pure cropping. In the mixed farming systems of the sahelian zone, increasing the legume component in the farming systems is important in order to increase the availability of fodder as source of livestock feed while increasing soil fertility. The rotation of legume with cereal will give more opportunity to increase the legume component in the present cropping system.

Despite the recognized need to apply chemical fertilizers for high yields, the use of fertilizers in West Africa is limited by lack of capital, inefficient distribution systems, poor enabling policies and other socio-economic factors. Cheaper means of improving soil fertility and productivity are therefore necessary. Cereal/legume rotation effects on cereal yields have been reported for the sahelian zone of West Africa (Bagayoko et al. 1996; Bationo et al. 1998; Klaij and Ntare 1995; Stoop and van Staveren 1981; Bationo and Ntare 1999).

Table 8.6 shows the effect of cowpea-millet rotation on millet grain and total biomass production. In a period of 3 years, there was an increase of about 3 t ha^{-1} of

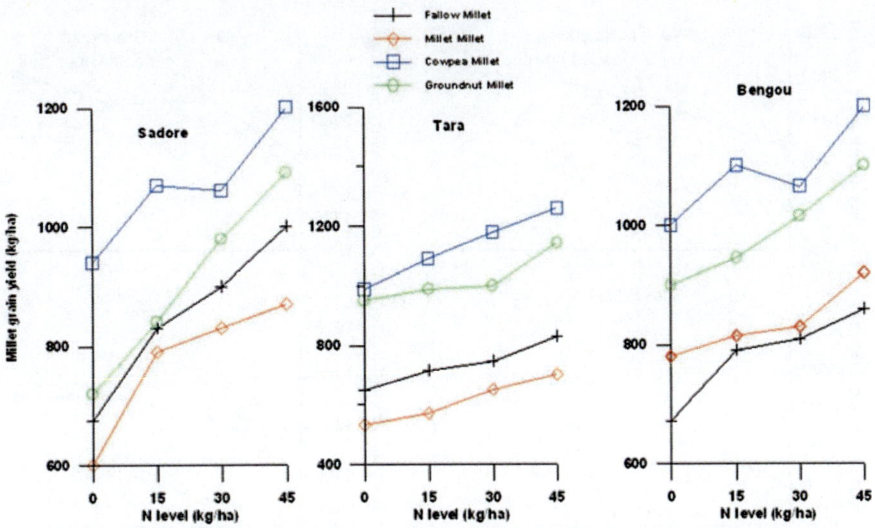

Fig. 8.4 Effect of nitrogen and rotation on pearl millet grain yield (Average of 4 years) at Sadore, Tara and Bengou

total dry matter production and about 0.65 t ha^{-1} of grain yield when millet was grown in rotation with cowpea.

At Sadoré, Niger, nitrogen use efficiency increased significantly when millet was rotated with cowpea as compared to continuous millet cultivation. At Sadoré in 1990, the nitrogen use efficiency increased from 20% in continuous pearl millet cultivation to 28% when pearl millet was rotated with cowpea. For both Bengou and Sadoré nitrogen derived from the soil was better used in rotation systems than with continuous millet. For example at Bengou in 1990, nitrogen derived from the soil increased from 39 kg Nha^{-1} in continuous pearl millet cultivation to 62 kg Nha^{-1} when pearl millet is rotated with groundnut. These clearly indicate that although all the above biomass of the legume will be used to feed livestock and not returned to the soil, rotation will increase not only the yields of succeeding cereal but also its nitrogen use efficiency.

Bationo and Ntare (1999) carried out a long-term experiment to investigate the effect of continuous monoculture as compared to crop rotation, N fertilizer application on crop yield and soil chemical properties. At all the three sites rotation of pearl millet with groundnut and cowpea resulted in higher significant pearl millet grain yields than in the monoculture cropping of pearl millet on average over the 4-year period (Fig. 8.4). Grain yield of millet following cowpea was higher than millet following groundnut at the three sites. Similarly, millet total dry matter was higher in the cowpea/millet rotation, except at Bengou where the total dry matter was higher in the groundnut/millet rotation system. With no application of N fertilizer millet grain yield after cowpea increased by 57% at Sadore, 18% at Bengou and 87% at Tara. Even at higher levels of N, continuous cropping of millet produced the lowest yields of both millet grain and total dry matter. The response of legumes to rotation with millet is

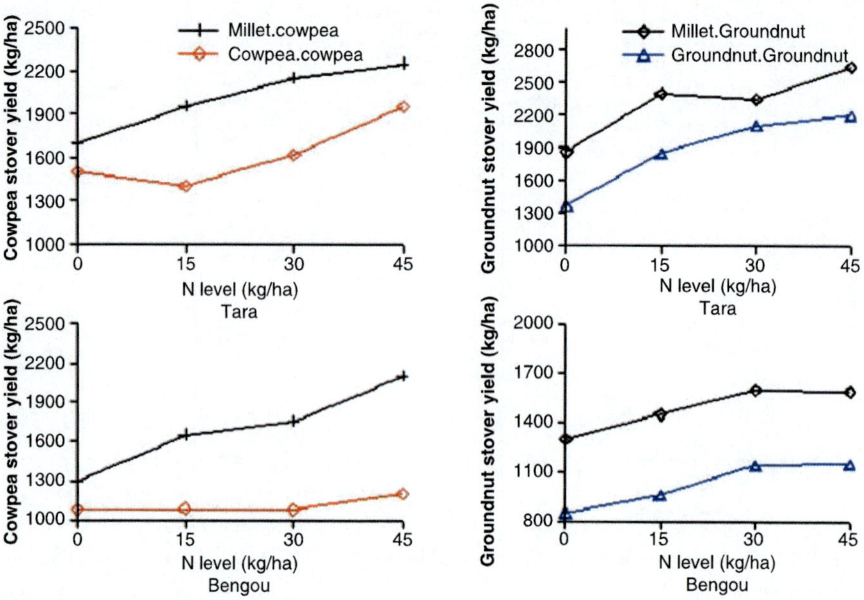

Fig. 8.5 Effects of nitrogen and rotation on legume stover yield (kg/ ha), average of 4 years at Tara and Bengou

also apparent at Tara and Bengou where legume yields were consistently lower in monoculture than when rotated with millet (Fig. 8.5). This suggests that factors other than N alone contributed to the yield increases in the millet-legume rotations. A common assumption by a number of workers has been that the positive effect of rotations arise from the added N from legume in the cropping system. Nonetheless, some workers have attributed the positive effects of rotations to the improvement of soil biological and physical properties and the ability of some legumes to solubilize occluded P and highly insoluble calcium bounded phosphorus by legume root exudates (Gardner et al. 1981; Arhara and Ohwaki 1989). Other advantages of crop rotations include soil conservation (Stoop and van Staveren 1981), organic matter restoration (Spurgeon and Grissom 1965) and pest and disease control (Sinnadurai 1973).

Bagayoko et al. (2000) in studies of cereal legumes effects on cereal growth in the sudano-sahelian zone of West Africa reported that the rotation effect although significant in most of the cases varied with sites and years. At Sadore as an example, whereas grain yield of pearl millet in 1998 was 1,557 kg/ha in the continuous millet production, the millet rotated with cowpea yielded 1,904 kg/ha (a 22% increase) and in same year, sorghum grain yield increased by 50% due to rotation with cowpea in the Sudanian zone (Table 8.7).

In order to estimate N availability, Bationo et al. (1999) incubated soils and determined mineral nitrogen at 7, 21 and 35 days (Keeney 1982). Crop rotation significantly affected mineral nitrogen (Fig. 8.6). The fallow millet rotation

Table 8.7 Millet and sorghum dry matter at thinning and grain and total dry matter yield at harvest as influenced by millet/cowpea cropping systems at Gaya, Goberi and Sadoré (Niger) and sorghum groundnut cropping systems at Kouaré (Burkina Faso)

Sites and Cropping system	Thinning					Harvest		
	TDM		Grain yield			TDM[b]		
	1996	1998	1996	1997	1998	1996	1997	1998
				kg ha^{-1}				
Sadoré								
Continuous millet	4	6	937	321	1,557	4,227	2,219	6,992
Millet after cowpea	5	7	1,255	340	1,904	5,785	2,832	8,613
P>F[a]	<0.001	0.059	<0.001	0.344	<0.001	<0.001	<0.001	<0.001
Goberi								
Continuous millet	13	8	–	779	956	2,328	3,444	4,220
Millet after cowpea	16	9	–	827	1,151	2,579	3,743	4,800
P>F	<0.001	0.063	–	0.145	<0.001	<0.001	0.019	<0.001
Gaya								
Continuous millet	13	13	794	378	645	2,601	2,469	2,598
Millet after cowpea	15	11	889	466	636	2,823	2,684	2,510
P>F	0.001	0.014	0.085	<0.001	0.801	0.025	0.069	0.353
Kouaré								
Continuous sorghum	30	26	397	786	238	3,056	4,505	2,689
Sorghum after cowpea	36	25	553	884	357	4,191	5,316	3,633
P>F	<0.001	0.352	<0.001	<0.001	<0.001	<0.001	<0.001	<0.001

[a]Probability of a treatment effect (significance level)
[b]Total dry matter

supplied more nitrogen than the legume millet rotation, but the latter was more productive for millet production. As suggested earlier, these results suggest that other factors in addition to biological nitrogen fixation may be involved in the positive effect of legume cereal rotation. Crop rotation is known to substantially increase soil microbial activity and this may lead to an increase in nutrient availability.

Results of field trials in sandy sahelian soils showed that P application has a very significant effect on yield of pearl millet grain and total dry matter yields in the different cropping systems (Figs. 8.7 and 8.8). Rotations of pearl millet performed better than the continuous cultivation. From the data in Fig. 8.9 it can be concluded also that cowpea yields significantly increased with the application of P fertilizers. When cowpea was grown after an intercrop of millet with cowpea or pure millet, the cowpea stover yields were higher than when cowpea preceded cowpea. Therefore, it could also be deduced that rotation has an effect on the legume. The data in Table 8.8 give the land equivalent ratio in different cropping systems. The land equivalent ratio varied from 24% to 200% indicating that even with the use of external input, intercropping is better than pure cropping. In addition, the application of N fertilizer with phosphorus did not affect much the land equivalent ratio. Other

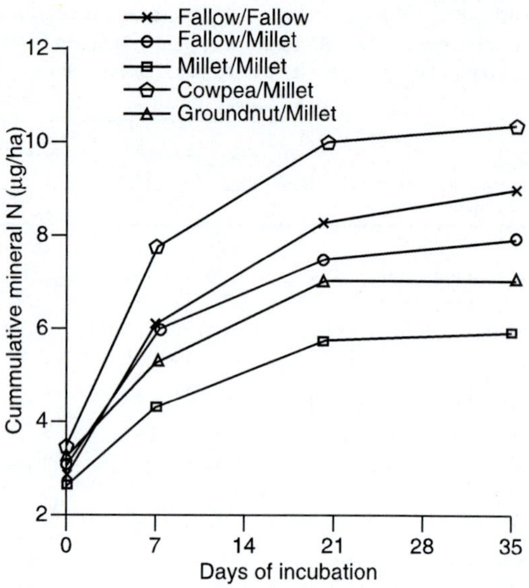

Fig. 8.6 Relationship between cumulative mineral nitrogen and time of incubation of soils from different crop rotations pooled over three sites

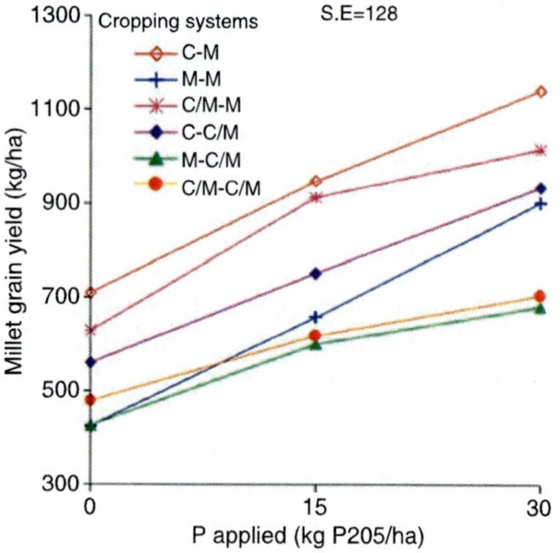

Fig. 8.7 Effect of phosphorus and nitrogen on millet grain yield under different cropping systems over 4 years, Sadore, Niger

investigations on rotation between pearl millet and improved fallow of *Stylosanthes pacheycarpa* gave similar results as rotation with groundnut (Fig. 8.10).

In the long-term cropping systems management trials, changes in soil chemical properties have been monitored in different agro-ecological zones of the sudano-sahelian region. As compared to adjacent fallow, rotations resulted in significantly lower soil pH, Effective Cation Exchange Capacity (ECEC) and base saturation at

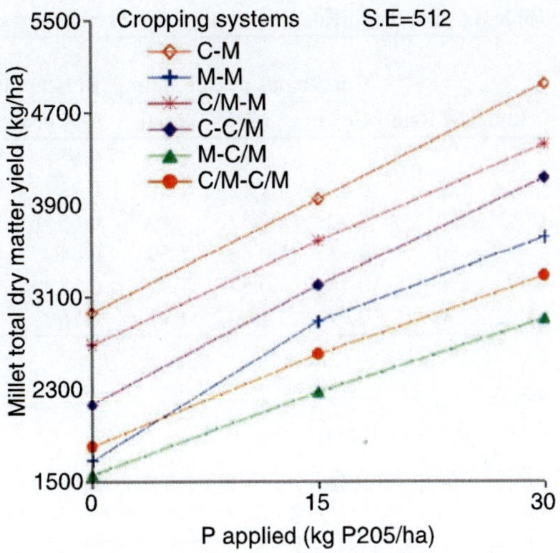

Fig. 8.8 Effect of phosphorus and nitrogen on millet total dry matter yield under different cropping systems over 4 years, Sadore, Niger

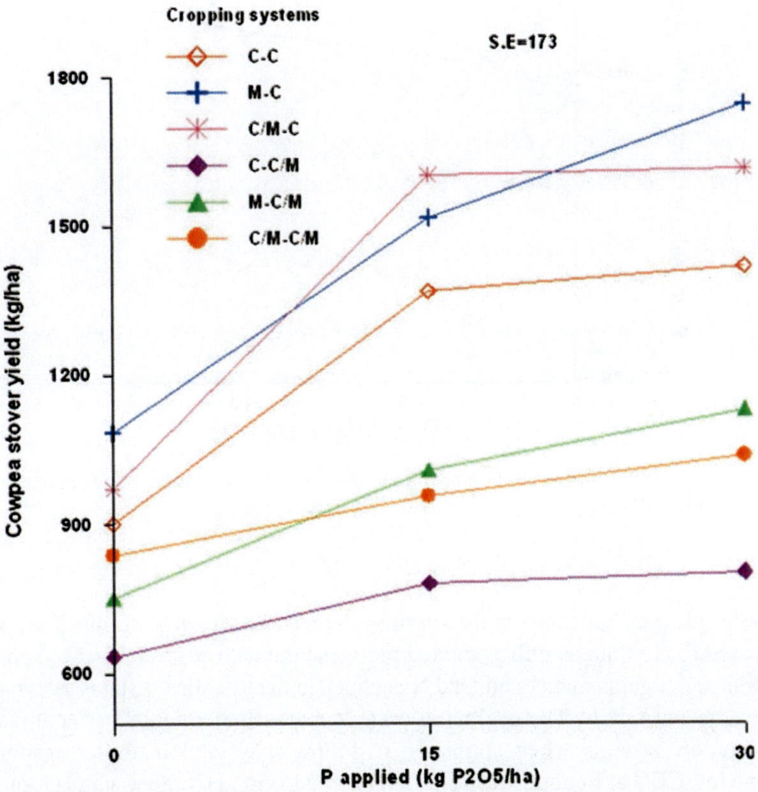

Fig. 8.9 Effect of phosphorus and nitrogen on cowpea stover yield under different cropping systems over 4 years, Sadore, Niger

Table 8.8 Land equivalent ratios in different cropping systems over a period of 4 years

		Continuous intercropping			Rotation following millet			Rotation following cowpea		
P. Rate	N Rate	Cowpea	Millet	Total	Cowpea	Millet	Total	Cowpea	Millet	Total
0	0	1.12	0.92	2.04	0.65	0.87	1.52	0.71	0.76	1.47
	30	0.85	1.12	1.97	0.72	0.85	1.57	0.93	0.84	1.77
15	0	0.77	0.72	1.49	0.65	0.69	1.34	0.58	0.73	1.31
	30	0.73	0.86	1.59	0.64	0.78	1.42	0.61	0.86	1.47
	0	0.76	0.57	1.33	0.57	0.67	1.24	0.57	0.83	1.40
30	30	0.77	1.00	1.77	0.76	1.06	1.82	0.48	0.88	1.36

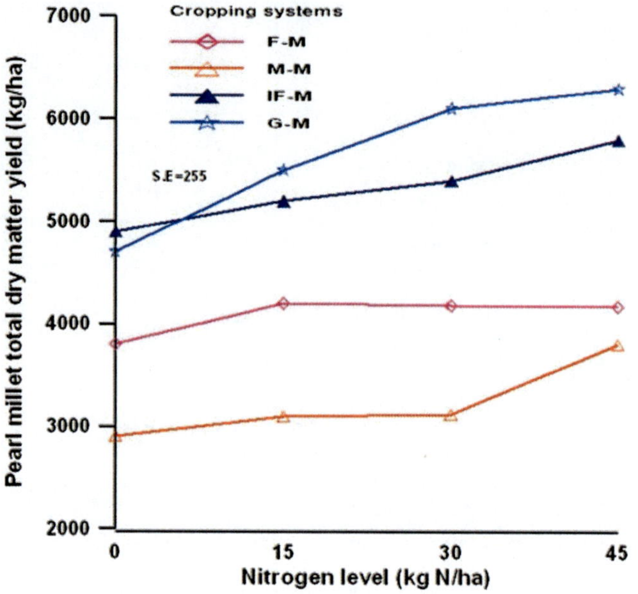

Fig. 8.10 Effect of nitrogen and different cropping systems on pearl millet total dry matter yield Bengou, Niger, 1994

Sadore and Tara but these were slightly changed at Bengou (Table 8.9). When compared to continuous millet, rotated plots maintained the same level of soil acidity. Both soil organic matter and total N declined in the rotations systems as compared to the adjacent fallow. The decline in organic matter under continuous cultivations and rotation systems when compared to fallow may explain the corresponding decline in ECEC at Bengou and Tara. This finding is in agreement with Bationo and Mokwunye (1991) who reported that in the sudano-sahelian zone, ECEC is more related to organic matter than to the clay content of the soils.

Table 8.9 Soil chemical properties after 5 years of experimentation at different sites in the Sudano-sahelian region

Rotation	pH			Organic matter (%)			Total N (mg kg^{-1})			Effective cation exchange capacity (cmol kg^{-1})			Base saturation (%)		
	Sadore	Bengou	Tara	Sadore	Bengou	Tara	Sadore	Bengou	Tara	Sadore	Bengou	Tara	Sadore	Bengou	Tara
Initial	4.1	4.3	4.1	0.22	0.20	0.45	74	226	197	0.54	1.87	1.20	57	89	58
F - F	4.7	4.7	5.0	0.76	0.46	0.56	351	230	219	1.95	1.15	1.25	96	94	97
F - M	4.9	4.7	5.0	0.74	0.44	0.59	302	251	2.07	1.83	1.35	1.16	98	93	94
M - M	4.6	4.4	4.3	0.52	0.37	0.44	235	178	165	1.91	1.11	0.88	95	83	81
C - M	4.7	4.3	4.3	0.56	0.35	0.47	260	206	197	1.84	1.15	0.88	97	80	76
G - M	4.6	4.3	4.2	0.58	0.27	0.45	263	192	130	1.88	1.25	0.81	93	80	70
S.E(D.F 27)	0.11	0.09	0.10	0.030	0.040	0.033	12.0	10.67	21.3	0.123	0.107	0.077	0.016	3.66	2.67
CV(%)	4	4	4	9	20	13	8	10	23	13	17	15	3	9	6

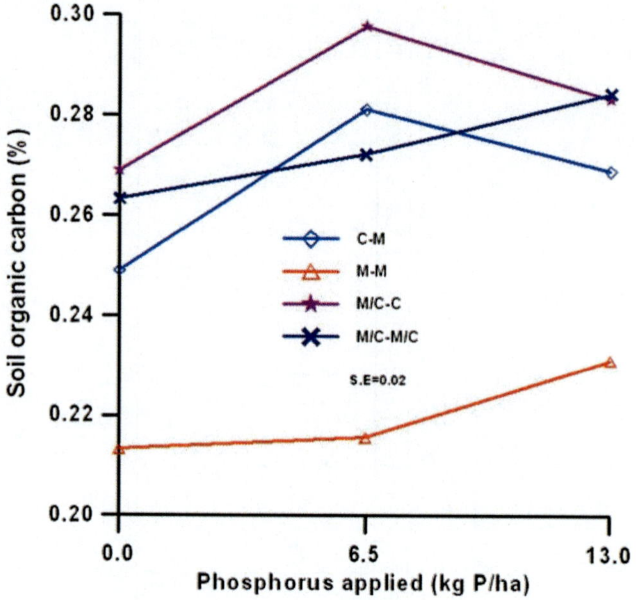

Fig. 8.11 Effect of phosphorus and cropping system on soil organic carbon, Sadore, Niger, 1995

Table 8.10 Analysis of variance expressed as percentage of total sun of square of soil chemical characteristics as affected by different cropping systems, nitrogen and phosphorus over a period of 5 years

	Degree of fee	Organic carbon	pH	Mineral N	Bray P	P soluble in water
Repetition	2	1.84	0.34	15.52	1.25	9.85
Cropping system (CS)	8	15.53**	11.4**	6.01*	1.71	1.20
Nitrogen (N)	2	0.45	35.5**	0.3	0.48	0.60
Phosphorus (P)	2	2.53*	0.3	0.38	62.73**	1.95*
CS × N	16	5.91	2.4	2.50	2.51	5.80
CS × P	16	6.68	2.7	4.88	2.96	5.85
N × P	4	1.79	01.5	1.88	1.74	0.58
CS × N × P	32	11.25	07.6	10.35	4.61	12.03
Residual	160	54.01	35.2	58.17	22.00	60.14
Total	2,492	100	100	100	100	100

*Significant at 5%
**Significant at 1%

In the long-term cropping systems management trial at Sadore, rotation systems were found to have higher levels of organic carbon as compared to the continuous cropping system (Fig. 8.11). This could partially be due to the fall of cowpea leaves. The data in Table 8.10 give some soil chemical properties after 5 years of cultivation for different cropping systems, N and P. The different cropping systems have significant effect on the soil organic carbon. The soil organic carbon level was 0.22% in the continuous systems whereas it increased to 0.27% in the rotation system. As a result of this, soil pH was higher in the rotation systems as compared to the continuous monoculture. The application of N fertilizers resulted in soil acidification. As expected mineral nitrogen was higher in the rotations involving cowpea as compared to the monoculture systems.

New Research Opportunities for ISFM in Sudano-Sahelian Zone of West Africa (SSZWA)

Integrated Soil Fertility Management (ISFM) has been viewed as the utilization of available organic and inorganic sources of plant nutrients in an efficient manner to improve on nutrient use efficiency by the growing crops. However, cropping systems are also forming an integral part of the ISFM. As discussed in the preceding sections of this paper, some cropping systems which have shown the potential for addressing the problem of land degradation in the SSZWA region have been researched and tested. Unfortunately, very few of these technologies have been adopted by the resource poor farmers. Therefore future research should focus more on problems driven by socio-economic factors and enabling policy environment in order to enhance farmers' capacity to invest in soil fertility restoration. The adoption of the participatory approach with farmers carrying out problem diagnosis, planning and implementing soil fertility restoration technologies and evaluating the technologies themselves will be essential. In this way, the technologies generated have a better chance of adoption by end users.

Conclusion

Intercropping and rotation systems have proved superior in several studies in the sudano-sahelian zone of West Africa, therefore, there is need to have these technologies tested together with farmers on-farm to increase efficiency in adoption. Farmer trials are indispensable in the scaling up of technologies and can be a good pillar in strengthening the ISFM approaches. Consideration of the economic benefits of the technologies will be a vital component for the success of these technologies. Thus markets are increasingly becoming part of the research process since farmers have to trade off cash crops and the excess of their food crop produce. Participatory farmer research involving produce prices, post harvest processing and storage, farmer cooperatives should be conducted and may become an important incentive for technology adoption.

References

Abalu GOI (1976) A note on crop mixtures under indigenous conditions in Northern Nigeria. J Dev Stud 12:11–20

Arhara J, Ohwaki Y (1989) Estimation of available phosphorus in vertisol and alfisol in view of root effects on rhizosphere soil. XI Colloquium, Wageningen

Bagayoko M, Mason SC, Traore S, Eskridge KM (1996) Pearl millet/cowpea cropping systems yield and soil nutrient levels. African Crop Sci J 4:453–462

Bagayoko M, Buerkert A, Lung G, Bationo A, Romheld V (2000) Cereal/legume rotation effects on cereal growth in Sudano-Sahelian West Africa: soil mineral nitrogen, mycorrhizae and nematodes. Plant Soil 218:103–116

Baker EFI (1979) Mixed cropping in Northern Nigeria. III, mixtures of cereals. Exp Agric 15:41–48

Baker EFI (1980) Mixed cropping in Northern Nigeria. IV. Extended trials with cereals and groundnuts. Exp Agric 16:361–369

Bationo A, Mokwunye AU (1991) Alleviating soil fertility constraints to increased crop production in West Africa. The experience in the Sahel. Fert Res 29:95–115

Bationo A, Ntare BR (1999) Rotation and nitrogen fertilizer effects on pearl millet, cowpea and groudnut yield and soil chemical properties in a sandy soil the semi arid tropics, West Africa. J Agric Sci, Cambridge 2000 134:000–000

Bationo A, Ntare BR (2000) Rotation and nitrogen fertilizer effects on pearl millet, cowpea and groundnut yield and soil chemical properties in a sandy soil in the semi-arid tropics, West Africa. J Agric Sci 134(3):277–284

Bationo A, Lompo F, Koala S (1998) Research on nutrient flows and balances in West Africa: state-of-the-art. Agric Ecosyst Environ 71:19–35

Bationo A, Henao JA, Kumar A (1999) Phosphorus use efficiency as related to P fertilizers, rainfall, soil and crop management in West African semi-arid tropics: phosphorus use efficiency in West Africa. In: Proceedings of the international workshop food security in nutrient-stressed environments: exploiting plants genetic capabilities, Patancheru, Andhra Pradesh, India. ICRISAT, Patancheru, Andhra Pradesh, India, and JIRCAS, Tsukuba, Ibaraki, Japan, 27–30 Sept 1999, p 20

Bationo A, Ntare BR, Tarawali S, Tabo R (2001a) Soil fertility management and cowpea production in the Semi-Arid tropics of West Africa. Paper for the world cowpea research conference III held at IITA, Ibadan, Nigeria, 4–7 Sept 2000. Presented in 2000

Bationo A, Ntare BR, Tarawali S, Tabo R (2001b) Soil fertility management and cowpea production in the Semi-Arid tropics of West Africa. Paper for the world cowpea research conference III held at IITA, Ibadan, Nigeria, 4–7 Sept 2000. Presented in 2000

Bationo A, Vanlauwe B, Kihara J, Kimetu J (2003) Use of mineral and organic fertilizers to increase land sustainability and productivity. CTA international seminar on information support for sustainable soil fertility management in October 2003

Bationo A, Kimetu J, Ikerra S, Kimani S, Mugendi D, Odendo M, Silver M, Swift MJ, Sanginga N (2004) The African network for soil biology and fertility: new challenges and opportunities. In: A. Bationo (ed.) Managing Nutrient Cycles to Sustain Soil Fertility in Sub-Saharan Africa. Academy Science Publishers, Nairobi, pp 1–23

Blumenthal MJ, Quach VP, Searle PGE (1982) Effect of soybean population density on soybean yield, nitrogen accumulation and residual nitrogen. Aust J Exp Agric 28:99–106

Finlay KW, Wilkinson GN (1963) The analysis of adaptation in a plant-breeding programme. Aust J Agr Res 14:742–754

Fussell LK (1985) Evaluation of millet/cowpea intercropping system in western Niger. In: Proceedings of the regional workshop on intercropping in the Sahelian and Sahelo-Sudanian zones of West Africa, Institut du Sahel, Niamey, Niger, Bamako, Mali, 7–10 Nov 1984, pp 26–39

Fussell LK, Serafini PG (1985) Associations de cultures dans les zones tropicales semi-arides d'Afrique de l'Ouest: strategies de recherché anterieures et futures. In: Ohm HW, Nagy JG (eds) Technologies appropriees pour les paysans des zones semi-arides de l'Afrique de l'Ouest. Purdue University, West Lafayette, pp 254–278

Fussell LK, Serafini PG, Bationo A, Klaij MC (1986) Management practices to increase yield and yield stability of pearl millet in Africa. Paper presented at the international pearl millet workshop held at ICRISAT, Patancheru, Andhra Pradesh, India

Gardner MK, Parbery DG, Barker DA (1981) Proteoid root morphology and function in legumes alnus. Plant Soil 60:143–147

ICRISAT (1984–1988) Annual Report, 1984 to 1988. Patancheru, A.P. 5023, India, ICRISAT

IER (1987–1990) Institut d'Economie Rurale. Rapport Annuel et Commissions Techniques Production Vivrieres et Oleagineuses. Program cooperatif ICRISAT/Mali, IER, Bamako, Mali

Keeney DR (1982) Nitrogen availability indices. In: Page AL, Millet RH, Keeney DR (eds) Methods of soils analysis, Second edition agronomy monograph No. 9. American Society of Agronomy, Madison, pp 711–730

Klaij MC, Ntare BR (1995) Rotation and tillage effects on yield of pearl millet (Pennisetum glaucum) and cowpea (Vigna unguiculata), and aspects of crop water balance and soil fertility in semi-arid tropical environment. J Agric Sci, Cambridge 124:39–44

Murwira HK, Kirchmann H (1993) Nitrogen dynamics and maize growth in a Zimbabwean sandy soil under manure fertilisation. Commun Soil Sci Plant Anal 24(17–18):2343–2359

Norman DW (1974) Rationalizing mixed cropping under indigenous conditions: the example of northern Nigeria. J Dev Stud 11:3–21

Ntare BR (1989) Intercropping morphologically different cowpea with pearl millet in a short season environment in the Sahel. Exp Agric 26:41–47

Prins WH (1993) The role of plant nutrients for sustainable food crop production in sub-Saharan Africa. Vereniging van Kunstmest Producenten (Dutch Association of Fertilizer Producers), Leidschendam, pp. x + 232pp

Reddy KC, Visser P, Buekner P (1992) Pearl millet and cowpea yields in sole and intercrop system, and their after-effects on soil and crop productivity. Field Crop Res 28:315–326

Rupela OP, Saxena MC (1987) Nodulation and nitrogen fixation in chickpea. In: Saxena MC, Sigh KB (eds) The chickpea. Commonwealth Agricultural Bureaux International and International Center for Agricultural Research in the Dry Areas, Farnham Royal, pp 191–206

Serafini PG (1985) Intercropping systems: the ICRISAT Mali experience: 1979–1983. In: Proceedings of the regional workshop on intercropping in the Sahelian and Sahelo-sudanian zones of West A1984, Institut du Sahel, Niamey, Niger, Bamako, Mali, pp 154–179

Shetty HR (1984) Crop volume and out-turn table for stands of Acacia nilotica. Indian Forester 110:586–590

Shetty SVR, Keita B, Sow AA, Serafini PG (1987) Improving millet-based cropping systems in the Sahelo-Sudanian zone of Mali. In: Proceedings of the international pearl millet workshop, International Crops Research Inst. for the Semi-Arid Tropics, Patancheru, A.P. (India), ICRISAT, p 306

Sinnadurai S (1973) Crop rotation to control nematodes in tomatoes. Ghana J Agric Sci 6:137–139

Sivakumar MVK (1988) Predicting rainy season potential from the onset of rains in Southern Sahelian and Sudanian climatic zones of West Africa. Agr Forest Meteorol 42:295–305

Sivakumar MVK, Renard C, Klaij MC Ntare BR, Fussel LK, Bationo A (1990) Natural resource management for sustainable agriculture in the Sudano-Sahelian zone. In: International symposium on Natural Resources Management for Sustainable Agriculture, New Delhi, India, 6–19 Feb 1990

Spurgeon WI, Grissom PH (1965) Influence of cropping systems on soil properties and crop production. Missisipi Agriculture Experiment Station Bulletin No. 710

Steiner KG (1984) Intercropping in tropical smalholder agriculture with special reference to West Africa. Stein, West Germany, 304 pp

Stoop WA, Staveren JPV (1981) Effects of cowpea in cereal rotations on subsequent crop yields under semi-arid conditions in upper-Volta. In: Graham PC, Harris SC (eds) Biological nitrogen fixation technology for tropical agriculture. Centro International de Agricultura Tropical, Cali

Swinton SM, Numa G, Samba LA (1984) Les cultures associees en milieu paysan dans deux regions du Niger: Filingue et Madarounfa. In: Proceedings of the Regional Workshop on Intercropping in the Sahelian and Sahelo-sudanian zones of West Africa, Institute du Sahel, Niamey, Niger, Bamako, Mali, 7–10 Nov 1984, pp 183–194

Tanaka N, Fujii H, Yoshida A (1983) Balance sheet of soybean nitrogen (in Ja). Nogyo Gijutsu 38:71–72

Willey RW (1979) Intercropping – its importance and research needs. Part 2. Agronomy and research approaches. Field crop Abstracts 32:733–785

Willey RW, Natarajan M, Reddy MS, Rao MR (1985) Intercropping to make use of limited moisture supply and to minimize the effects of drought. In: La secheresse en zone intertropicale pour une lutte integree. CIRAD Boudin Press, Paris, pp 421–429, 591 pp

Chapter 9
Agronomic and Economic Evaluation of Fertilizer Use in West Africa from Long-Term Experiments: Alternative Phosphorus Fertilizer Sources in Different Agro-Ecological Zones of Togo

A. Bationo, A.K. Pinto-Toyi, E. Ayuk, and A.U. Mokwunye

Abstract Phosphorus deficiency is a major constraint to crop production in West Africa. However, the use of P fertilizers is limited by both the unavailability and the high cost of imported commercial fertilizers.

Most of the soils in West Africa, in addition to low phosphorus content, have low buffering capacities as well as a low P retention capacity. The trials, conducted in the humid and sub-humid zones of Togo showed that P fertilizers as well as the local phosphate rock can be used to improve crop production. The results also showed that the Residual Effectiveness Index (REI) of all the P sources remained higher than 60% even after 3 years of cropping. This quantification of the "legendary" high residual effectiveness of P fertilizers in West African soils could lead to the development of fertilizer management systems working in partnership with limited-resource farmers. In spite of the excellent performance of the local phosphate rock, the P source with the highest net gain over the trial period was single superphosphate (SSP). Evaluation of the Relative Agronomic Effectiveness of the P sources using SSP as the standard showed that SSP was by and large the superior

A. Bationo (✉)
Soil health, Alliance for a Green Revolution in Africa (AGRA), Accra, Ghana
e-mail: abationo@agra-alliance.org

A.K. Pinto-Toyi
IFDC, Lome, Togo

E. Ayuk
UNU, Accra, Ghana

A.U. Mokwunye
Consultant, Accra, Ghana

A. Bationo et al. (eds.), *Lessons learned from Long-term Soil Fertility Management Experiments in Africa*, DOI 10.1007/978-94-007-2938-4_9,

P source. A change in paradigm with regards to the role of local phosphate rocks in crop production is needed. The data from these trials definitely suggested that local phosphate rock could be used to improve the P fertility of the soils. Therefore, instead of evaluating the local phosphate rock in terms of its Relative Agronomic Effectiveness (RAE), consideration should be given to examining the use of the rock to recapitalize the soil fertility. Sustainable cropping systems in West Africa depend on the long-term improvement of the capital pool of phosphorus in the soils. It is suggested that West African governments take the lead in the utilization of the abundant phosphate rock deposits in the region as amendments to improve the productivity of the soils on a long-term basis.

Keywords Phosphorus deficiency • P fertilizer • Residual effectiveness index (REI) • West Africa • Phosphate rock

Introduction

Several trials carried out on various crops in tropical Africa have shown that lack of phosphorus (P) and nitrogen (N) constitute major constraints to food crop production (Milne 1968; Richardson 1968; Pichot and Roche 1972).

According to Jones and Wild (1975), P deficiency occurs widely in the humid and sub-humid zones of what is considered the southern and northern Guinea savanna zone. Early work identifying the extent of phosphorus deficiency in soils of Togo have been reviewed. Similar low quantities of both total and available P have been reported for soils of the West African savanna regions. West Africa is endowed with several deposits of phosphate rock (PR) and there have been several efforts in the past to evaluate the agronomic performance of the finely ground phosphate rock from these local deposits (Nabor et al. 1974; Jenny 1973; Jones 1973; Mokwunye 1979; Thibout et al. 1980). These trials were driven by the desire to find "cheaper alternatives" to imported commercial superphosphates. The use local PR for "corrective fertilization" involved a one-time application of a large dose of the PR (basal dose) followed by smaller annual "maintenence" doses. The corrective dose was aimed at rectifying the inherent poor nutrient status of the soils. The maintenence doses replaced what the crops removed each year. This practice was expected to take advantage of the reputed high residual effects of P fertilizers in West African soils. Unfortunately, no attempt has ever been made to quantify this residual effectiveness. Quantifying the residual effectiveness of various phosphorus fertilizer sources applied to West African soils would facilitate the design of fertilizer management practices that would take advantage of residual fertilizers and be more ideal for the limited resource farmers who cannot purchase P fertilizers every year. The objectives of these trials were to monitor the performance of Togo phosphate rock applied to maize in both the humid and sub-humid regions of Togo and to attempt to quantify the residual effectiveness of various phosphorus sources. It was also important to evaluate the profitability of the use of the phosphorus sources for maize production in Togo.

Materials and Methods

Location and Treatments Tested

There are two principal agroecological zones in Togo. These are defined primarily by the rainfall distribution pattern. IFDC-Africa, the Africa Division of the International Fertilizer Development Center (IFDC) has set up trials in five locations covering the two agroecological zones. The five experimental sites are Amoutcho, Davié, Kaboli, Kokoumbo and Tchitchao. Davié and Amoutchou have bimodal rainfall pattern, while the rainfall pattern is unimodal in Kaboli, Tchitchao and Koukoumbo. The soils are predominantly Alfisols with associated Ultisols. The chemical and physical characteristics of the soils are presented in Table 9.1. These soils are largely coarse textured and are characteristically low in organic matter, total nitrogen and effective cation exchange capacity. They are deficient in available phosphorus and except for the plinthustalf at Kaboli the maximum P sorbed is also low.

The P fertilizer sources tested were finely ground Hahotoe (Togo) phosphate rock (PR), Togo PR acidulated with sulfuric acid at 50% (PAPR 50) and Single Superphosphate (SSP). Some of the chemical characteristics of Togo PR and PAPR50 are presented in Table 9.2. The PAPR50 was produced at IFDC"s pilot plant using 50% of the sulfuric acid required to convert Togo PR to SSP as calculated from the composition of the rock. The size of PAPR50 granules ranges from 1.18 mm to 3.35 mm (minus 6 plus 14 mesh).

Two P fertilizer management strategies were adopted. SSP and PR were applied either annually or once every 3 years. The annual rates applied for either source were 0, 8.8, 17.6 and 35.2 kg P/ha. In addition to the annual application of PR (PRA) at the rates already stated some plots received PR once every 3 years (PRB) at the rates of 0, 26.4, 52.8 and 105.6 kg P/ha.

Nitrogen was applied to all plots each year at the rate of 90 kg N/ha as urea. N was split-applied with half at 15 days after planting and the other half at 45 days after sowing.

Potassium was applied each year to all plots before planting at the rate of 49.8 kg K/ha as muriate of potash.

The trials were carried out for three cycles of 9 years at Amoutchou from 1987 to 1995, two cycles of 6 years at Davié and Koukoumbo from 1989 to 1995 and one cycle each of 3 years at Kaboli (1991 to 1992) and Tchitchao (1987 to 1989).

Table 9.1 Soil taxonomy, texture and selected chemical properties of experimental sites

	DAVIÉ		AMOUTCHOU		TCHITCHAO		KABOLI		KOKOUMBO	
	0–20 cm	20–40 cm	0–20 cm	20–40 cm	0–20 cm	20–40 cm	0–20 cm	20–40 cm	0–20 cm	20–40 cm
Taxonomy	Kandiustalf		Paleustalf		Plinthustalf		Plinthustalf		Kanhaplustult	
Clay (%)	15	27	3.4	3	6	30	16	42	5	12
Sand (%)	77	66	83	82	81	59	76	50	88	79
Soil pH (H_2O)	6.03	5.91	6.12	6.00	5.71	5.38	5.87	5.5	5.92	6.01
Soil pH (KCl)	5.10	5.02	4.76	5.24	4.24	4.04	4.84	4.6	5.08	4.97
Organic matter (%)	0.81	0.49	1.08	267	0.3	0.60	1.09	0.82	0.44	0.24
Total nitrogen (ppm)	455	225	558	1.16	228	36	550	454	245	192
Available P (Bray P1) mg kg^{-1}	1.38	0.71	3.97	1.20	0.59	0.24	4.57	0.36	0.83	0.47
Exchangeable bases (Cmol kg^{-1})										
Ca	1.72	1.48	2.84	1.49	0.55	2.00	1.30	1.72	1.34	0.81
Mg	0.77	1.21	0.49	0.36	0.22	0.83	0.50	0.69	0.28	0.45
K	0.16	0.09	0.26	0.15	0.10	0.16	0.38	0.30	0.17	0.19
Na	0.09	0.14	0.09	0.13	0.10	0.14	0.09	0.27	0.10	0.09
Exchangeable acidity (Cmol kg^{-1})	0.04	0.02	0.03	0.03	0.34	0.52	0.10	0.34	0.04	0.04
Effective cation exchange capacity (Cmol kg^{-1})	2.78	2.94	3.71	2.16	1.21	3.65	2.37	2.98	1.93	1.58
Al Saturation	0.72	0.00	0.54	0.92	12.40	9.59	2.53	12.40	1.04	1.27
Maximum P sorbed (mg kg^{-1})	46	ND	23	ND	41	ND	183	ND	30	ND
P fertilizer equivalent (mg kg^{-1})[a]	1.75	2.10	1.31	1.34	1.66	4.36	3.38	11.62	1.71	2.12

Water extractable phosphorus (mg kg^{-1})[b]	1.23	0.60	1.16	0.86	0.53	0.24	0.71	0.00	1.16	0.49
Oxalate (mg kg^{-1})										
Al	357	449	158	147	196	614	579	1,122	120	178
Fe	607	920	419	310	345	730	1,452	1,773	344	441
Dithionite (mg kg^{-1})										
Al	238	353	214	242	310	621	442	701	273	387
Fe	577	1,063	435	302	685	1,377	1,288	1,156	300	668

ND Not determined

[a]Quantity of P need to increase Bray P1 by 1 mg kg^{-1}

[b]Water extractable phosphorus after addition of 25 mg P kg^{-1}

Table 9.2 Chemical analysis[a] of Togo PR and PAPR's (%)

Items	Togo PR	Togo PAPR
Total P (%)	15.7	11.0
Water-soluble P	0	5.1
Citrate soluble P_b	1.31	0.9
CaO	51.3	NA
Fe	4.0	NA
Al_2O_3	1.1	NA
Fe_2O_3	1.3	NA
MgO	0.05	NA
Na_2O	0.23	NA
K_2O	0.04	NA
SiO_2	0.03	NA
CO_2	4.6	NA

NA Not available
[a]Analysed by the International Fertilizer Development Center (IFDC)
[b]Citrate soluble P measured in neutral ammonium solution

A split plot design was used with five replications at all sites. Maize, variety Pozarica was grown in plot of 27 m^2 and with 75 cm between rows and 25 cm within rows.

Results and Discussion

Agronomic Evaluation

The results of the analysis of variance of the effect of different sources, rates, and P fertilizer management strategies for the five sites are presented in Table 9.3. At all sites, the addition of phosphorus from each of the three sources increased maize stover and grain yields. Maize also responded to varying rates of P. In most cases, there was a significant year effect, and, a significant interaction between years, P rates and P sources were observed.

For all sites except at Kaboli (grain yield), there was a significant P management strategy effect of P fertilizers. The application of the local phosphate rock once every 3 years (corrective fertilization) resulted in higher gross yields over the 3 year period as compared to when the phosphate rock was applied annually. It would seem that the addition of the large dose of P as PR did not only improve the capacity of the soil to supply P. There is a need to investigate the use of large doses of PR as amendment to improve the overall fertility of the soil.

Figure 9.1 shows the relationships between the yields obtained at each rate of P for the three P sources when applied annually. The data clearly indicate that SSP out performed the two other P sources. The relative agronomic effectiveness (RAE) of the

Table 9.3 Combined analysis of variance expressed as percentage of total sum of squares for grain and stover yield, Togo

Source of	Davié			Amoutchou			Kaboli			Tchitchao			Koukoumbo		
variance	DF	Grain	Stover	DF	Grain	Stover	DF	Grain	Stover	DF	Grain	Stover	DF	Grain	Stover
Replication	4	6.4**	5.2**	4	0.1 NS	1.8*	4	5.3**	4.9NS	4	0.6NS	7.4**	4	1.6NS	3.1**
Years (Y)	5	44.8**	62.3**	8	33.3**	45.7**	2	35.6**	49.4**	2	2.4*	4.3**	5	5.3**	45.8**
Error	20	3.1	2.6	32	5.4	5.2	8	1.4	4.6	8	1.5	1.7	20	3.0	1.7
P source (PS)	3	2.3**	2.0**	3	10.5**	8.4**	3	7.0**	5.6NS	3	2.7**	3.1*	3	8.6**	5.1**
Y * PS	15	1 NS	1.1 NS	24	3.7**	2.3 NS	6	3.7**	1.0 NS	6	4.2**	3,1 NS	15	0.9 NS	0.9 NS
Error	72	4.9	3.5	108	3.9	4.0	36	5.4	3.4	36	4.8	9.4	72	15.2	8.8
P - Rate (PR)	3	14.8**	5.3**	3	16.8**	9.0**	3	14.7**	7.3**	3	50.5**	28.8**	3	15.9**	6.7**
Y * PR	15	4.2**	1.0**	24	0.9**	0.9**	6	0.5 NS	0.3 NS	6	1.1**	0.7 NS	15	2.9**	0.8**
PS * PR	9	1.3**	1.1**	9	4.0**	3.9**	9	3.9**	3.1**	9	1.8**	2.1**	9	4.1**	2.6**
Y * PS * PR	45	1.9**	1.7**	72	2.3**	1.5**	18	1.7*	0.9 NS	18	2.5**	2.3 NS	45	1.5 NS	1.2 NS
Strategy (S)	1	0.3**	0.1**	1	0.9**	0.9**	1	0.1 NS	0.3*	1	3.7**	3.2**	1	2.5**	1.4**
Y * S	5	0.5**	0.3**	8	0.4**	0.2*	2	0.6**	0.3 NS	2	0.9**	0.7*	5	1.4**	1.1**
PS * S	3	0.2**	0.1 NS	3	1.3**	0.7**	3	0.3 NS	0.2 NS	3	1.2**	1.1**	3	0.3 NS	0.2 NS
Y * PS * S	15	0.2 NS	0.2 NS	24	0.4 NS	0.8**	6	0.1 NS	0.2 NS	6	0.9**	1.3*	15	1.3*	0.7*
PR * S	3	0.2**	0.1 NS	3	0.3**	0.4**	3	0.06 NS	0.2 NS	3	1.6**	1.1**	3	1.1**	0.6**
Y * PR * S	15	0.3 NS	0.3 NS	24	0.3 NS	0.3 NS	6	0.3 NS	0.3 NS	6	0.5 NS	0.5 NS	15	0.8 NS	0.6 NS
PS * PR * S	9	0.3 NS	0.2 NS	9	0.5**	0.3**	9	0.3 NS	0.3 NS	9	0.7 NS	0.5 NS	9	1.2**	0.4 NS
Y * PS * PR * S	45	0.4 NS	0.4 NS	72	0.5 NS	0.5 NS	18	0.4 NS	0.4 NS	18	1.0 NS	0.9 NS	45	1.3 NS	1.0 NS
Error	672	12.8	12.8	1,008	14.4	13.4	336	18.6	17.5	336	17.4	27.9	672	31.1	17.2
TOTAL	959	100	100	1,439	100	100	479	100	100	479	100	100	959	100	100

NS Non significant
*Significant at 5%
**Significant at 1%

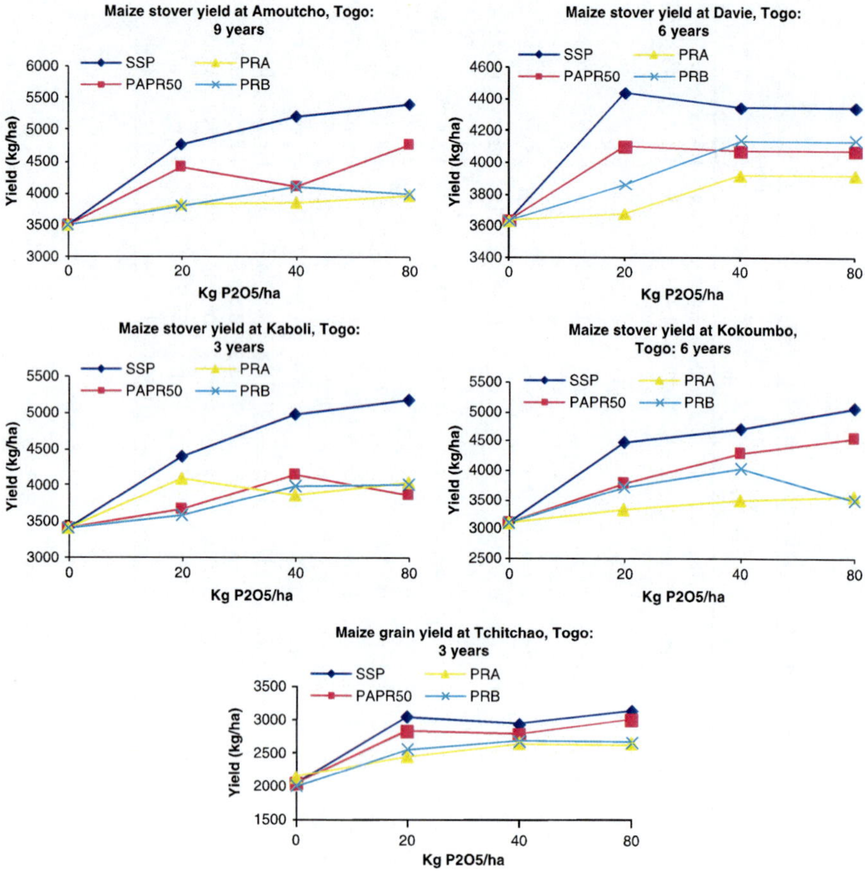

Fig. 9.1 Effect of rates and sources of P on maize stover yield in different agro ecological zones of Togo

phosphate rock and its partially acidulated form using SSP as standard was computed. The results presented in Table 9.4 show that: (1) The RAE ranged from 65% to 90%. The exception was at Kaboli where the response to P was generally poor. This is accounted for by the fact that Kaboli has the highest value of available P (Table 9.1). At all rates, improving the water soluble P content of PR through partial acidulation improved the agronomic effectiveness. (2) The highest RAE values for PR and PAPR were obtained at Tchitchao. In contrast with Kaboli, this Plinthustalf at Tchitchao had the lowest value of available P. The high concentration gradient between the PR or PAPR granule sites and the soil solution might have encouraged increased dissolution of the less soluble phosphate rock. (3) The difference between the performances of PR and PAPR was at Amoutchou. The very sandy Paleustalf at Amoutchou contains relatively high organic matter. Organic matter is a major source of available P in West African soils (Acquaye and Oteng 1972). The higher concentration of P in soil solution

Table 9.4 Relative agronomic effectiveness (RAE*) over years

		DAVIE				AMOUTCHOU				KABOLI				TCHITCHAO				KOUKOUMBO			
		Strategy I		Strategy II		Strategy I		Strategy II		Strategy I		Strategy II		Strategy I		Strategy II		Strategy I		Strategy II	
		Grain	Stover	Grain	Stover	Grain	Stover	Grain	Stover	Grain	Stover	Grain	Stover	Grain	Stover	Grain	Stover	Grain	Stover	Grain	Stover
P Source	P Rate																				
PAPR	20	57	62	19	54	80	75	74	69	44	47	20	3	86	81	89	83	63	44	71	57
	40	77	76	60	46	75	46	50	20	35	49	44	43	91	85	91	65	68	71	82	79
	80	73	67	69	46	87	73	68	58	31	28	30	22	91	82	90	83	84	93	55	49
Average		69	68	49	49	81	65	64	49	37	41	31	23	89	83	90	77	72	69	69	62
PRA	20	30	3	16	9	20	23	33	30	56	53	74	92	53	38	89	57	23	15	8	19
	40	56	36	56	45	17	19	24	22	36	33	27	23	59	69	63	61	52	36	2	1
	80	71	51	59	34	30	21	39	28	32	39	41	33	70	49	85	75	43	34	12	8
Average		52	30	44	29	22	21	32	27	41	42	47	49	61	52	79	64	39	28	7	9
PRB	20	21	13	52	47	28	21	61	27	34	16	77	21	63	48	86	71	58	42	60	49
	40	76	77	83	63	32	27	52	44	66	35	80	41	75	84	93	91	48	51	93	72
	80	86	84	91	76	34	26	40	25	62	31	85	39	72	53	94	75	37	23	24	15
Average		61	58	75	62	31	25	51	32	54	27	81	34	70	62	91	79	48	39	59	45

$$*RAE = \frac{\text{YEILD OF P - SOURCE} - \text{YEILD OF CONTROL}}{\text{YEILD OF SSP} - \text{YEILD OF CONTROL}} \times 100$$

at Amoutchou adversely affected the dissolution of PR. The agronomic results demonstrate that there are major differences in the way PR behaves in the different soils in Togo. The use of modeling is recommended in order to predict the effectiveness of PR in the different soils in Togo. This would make it possible to readily delineate areas where the direct application of the phosphate rock can be practised and areas where the processed form such as PAPR can be used for more effective performance.

The use of the two P fertilization management strategies in these facilitated the quantification of the residual effectiveness of the three P sources in the different soils. As indicated earlier, the residual effectiveness index (REI) was measured by comparing the yields of the previously fertilized plots with the yields of plots to which a fresh application of SSP was made during that season. The data in Table 9.5 show that even after three cropping seasons, the residual effectiveness indexes for all P sources were greater than 60%. What this means in practical terms is that a farmer in Davie who applies SSP at the recommended rate (17.6 kg P/ha) in any 1 year but is unable to reapply the fertilizer the following year would expect to maintain his yield, all other conditions favoring the good growth of maize in that year. In a region of the world where the high cost of P fertilizers limits their use, knowledge of the residual effectiveness of P fertilizers would facilitate the development of fertilizer management strategies that would take into account both the need for sustainable use of the land and the farmers capacity to purchase inputs. The high REI values is attributable to the inability of the West African soils to immobilize large amounts of soluble phosphates. All P sources gave residual effectiveness indexes (REI) values of more than 60% even after 3 years of P application. Except for Kaboli the maximum P adsorption was low for the other sites and as a result of the low P retention capacity, high residual effect of P and relatively small quantities of P fertilizers will be required for optimum crop production as compared to oxisols and ultisols found in more humid tropical regions. In Sierra Leone, only a maximum of 20% of the applied P is used by the first crop. By using P^{32} on the sandy Sahelian soils only 22% of P will be utilized the first year of cropping by pearl millet. When P is applied the second year on the same plot, the fertilizer P utilization decreased to 14% due to the residual effect of previous P application.

The data in Table 9.5 also indicate that there were differences between sites and sources for the REI values but because of limited number of observation it was not possible to make concrete conclusion about the physico-chemical factors responsible for those differences.

Stability Analysis

The relation between the environment which is the mean yield performance of the different treatments are reported in Table 9.6 and illustrated by Fig. 9.2 and Fig. 9.3. For both maize and stover yield and for the two management strategies, the control treatment performed worse than the other treatments in poor or good environment.

Table 9.5 Residual effectiveness index (REI*) for different phosphorus fertilizers sources at different agro-ecological zones of Togo

		1988		1989		1990		1991		1992		1993		1994		1995	
Sites	P-Sources	Grain	Stover	Grain	Stover	Grain	Stover	Grain	Stover	Grain	Stover	Grain	Stover	Grain	Stover	Grain	Stover
DAVIE	SSP	NA	NA	NA	NA	100	99	96	95	NA	NA	94	95	95	96	NA	NA
	PAPR	NA	NA	NA	NA	93	92	90	94	NA	NA	84	84	89	85	NA	NA
	PRA	NA	NA	NA	NA	88	87	85	87	NA	NA	84	87	85	87	NA	NA
	PRB	NA	NA	NA	NA	96	93	96	95	NA	NA	95	94	94	89	NA	NA
AMOUTCHOU	SSP	88	86	80	80	NA	NA	91	89	85	87	NA	NA	85	83	90	97
	PAPR	67	79	76	73	NA	NA	79	82	69	68	NA	NA	78	67	80	87
	PRA	68	74	85	79	NA	NA	72	82	64	63	NA	NA	76	65	73	86
	PRB	63	78	79	79	NA	NA	84	88	72	67	NA	NA	82	66	80	83
KABOLI	SSP	NA	NA	NA	NA	NA	NA	NA	NA	90	95	85	83	NA	NA	NA	NA
	PAPR	NA	NA	NA	NA	NA	NA	NA	NA	74	76	70	76	NA	NA	NA	NA
	PRA	NA	NA	NA	NA	NA	NA	NA	NA	77	83	76	77	NA	NA	NA	NA
	PRB	NA	NA	NA	NA	NA	NA	NA	NA	104	76	79	81	NA	NA	NA	NA
TCHITCHAO	SSP	68	101	66	74	NA	NA	NA	NA	NA	NA	NA	NA	NA	NA	NA	NA
	PAPR	66	94	74	74	NA	NA	NA	NA	NA	NA	NA	NA	NA	NA	NA	NA
	PRA	61	80	79	72	NA	NA	NA	NA	NA	NA	NA	NA	NA	NA	NA	NA
	PRB	75	82	97	76	NA	NA	NA	NA	NA	NA	NA	NA	NA	NA	NA	NA
KOUKOUMBO	SSP	NA	NA	NA	NA	90	85	80	76	NA	NA	87	94	78	81	NA	NA
	PAPR	NA	NA	NA	NA	87	83	72	70	NA	NA	72	71	68	63	NA	NA
	PRA	NA	NA	NA	NA	72	70	62	60	NA	NA	68	78*	64	66	NA	NA
	PRB	NA	NA	NA	NA	85	83	77	69	NA	NA	73	71	69	69	NA	NA

NA Not applicable $*REI = \frac{\text{YEILD OF TREATMENT}}{\text{YEILD OF FRESHLY APPLIED SSP}} \times 100$

Table 9.6 Linear regression equation of grain and stover yield of the environmental mean (EM) by treatment, Togo, 1987–1995

	Gain yield Strategy I			Grain yield Strategy II			Stover yield Strategy I			Stover yield Strategy II		
Treatment	Equation	R	Significance	Equation	R	Significance	Equation	R	Significance	Equation	R	Significance
Control	Y = −582 + 0.92 EM	0.84	**	Y = −443 + 0.95 EM	0.91	**	Y = −516 + 0.96 EM	0.93	**	Y = −345+0.98 EM	0.96	**
20 SSP	Y = 570 + 0.89 EM	0.79	**	Y = −394 + 0.93 EM	0.86	**	Y = 356 + 0.93 EM	0.86	**	Y = −323+0.95 EM	0.91	**
40 SSP	Y = 542 + 0.89 EM	0.79	**	Y = 33 + 0.88 EM	0.77	**	Y = 318 + 0.92 EM	0.84	**	Y = 167+0.92 EM	0.85	**
80 SSP	Y = 605 + 0.87 EM	0.76	**	Y = 519 + 0.91 EM	0.83	**	Y = 632 + 0.90 EM	0.82	**	Y = 380+0.95 EM	0.90	**
20 PAPR	Y = 357 + 0.87 EM	0.76	**	Y = 102 + 0.89 EM	0.80	**	Y = 469 + 0.93 EM	0.86	**	Y = −119+0.94 EM	0.88	**
40 PAPR	Y = 541 + 0.87 EM	0.76	**	Y = 135 + 0.91 EM	0.82	**	Y = 171 + 0.92 EM	0.86	**	Y = 149+0.92 EM	0.84	**
80 PAPR	Y = 605 + 0.83 EM	0.69	**	Y = 681 + 0.86 EM	0.74	**	Y = 10 + 0.91 EM	0.83	**	Y = 402+0.92 EM	0.85	**
20 PRA	Y = −277 + 0.94 EM	0.88	**	Y = −17 + 0.92 EM	0.84	**	Y = −46 + 0.96 EM	0.93	**	Y = −199+0.96 EM	0.92	**
40 PRA	Y = −159 + 0.89 EM	0.80	**	Y = 213 + 0.90 EM	0.82	**	Y = −4. + 0.94 EM	0.88	**	Y = −41+0.94 EM	0.89	**
80 PRA	Y = 66 + 0.90 EM	0.81	**	Y = 286 + 0.86 EM	0.75	**	Y = −80 + 0.97 EM	0.94	**	Y = 284+0.92 EM	0.85	**
60 PRB	Y = −265 + 0.93 EM	0.86	**	Y = −182 + 0.94 EM	0.88	**	Y = −90 + 0.97 EM	0.94	**	Y = −44+0.96 EM	0.91	**
120 PRB	Y = 28 + 0.88 EM	0.78	**	Y = 330 + 0.87 EM	0.76	**	Y = 369 + 0.92 EM	0.85	**	Y = 465+0.92 EM	0.84	**
240 PRB	Y = −336 + 0.88 EM	0.77	**	Y = 64 + 0.81 EM	0.66	**	Y = 16 + 0.92 EM	0.85	**	Y = 259+0.92 EM	0.85	**

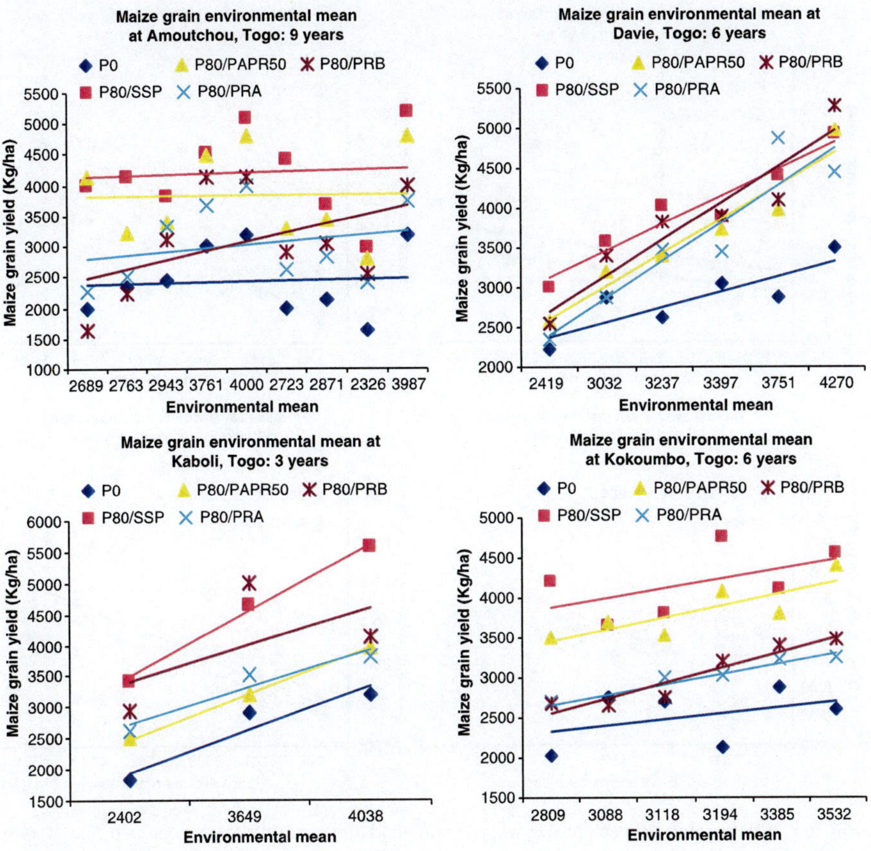

Fig. 9.2 Relationship between Maize grain yield and environmental mean in different agro ecological zones of Togo

In order to investigate if there was an interaction between the treatment and environment, the individual slopes reported in Table 9.6 were compared with the simple model which has a single slope for all treatments. For the first strategy, the coefficients of determination were 83% and 84% and respectively for the case of different slopes and the case of the same slope for grain yield and 82% for the stover yield. For the second strategy the coefficients of determination were 88% and 89% and, respectively, for the case of the different slopes and the case of the same slope for grain yield, and 88% and 89%, respectively, for the stover yield. The coefficient of determination obtained are quite very similar and we can conclude that the simple model with single slope is adequate. In other words, with the environment indexed by their mean yield, there is no evidence that any particular treatment performs differently from the others in poor or good environment. Nethertheless, it should be noted that the treatments and environment interaction can sometimes be a more complex relation that cannot be modeled in this simple way. One problem with the large number of treatments considered in this study is that isolated effects from particular treatments can be masked.

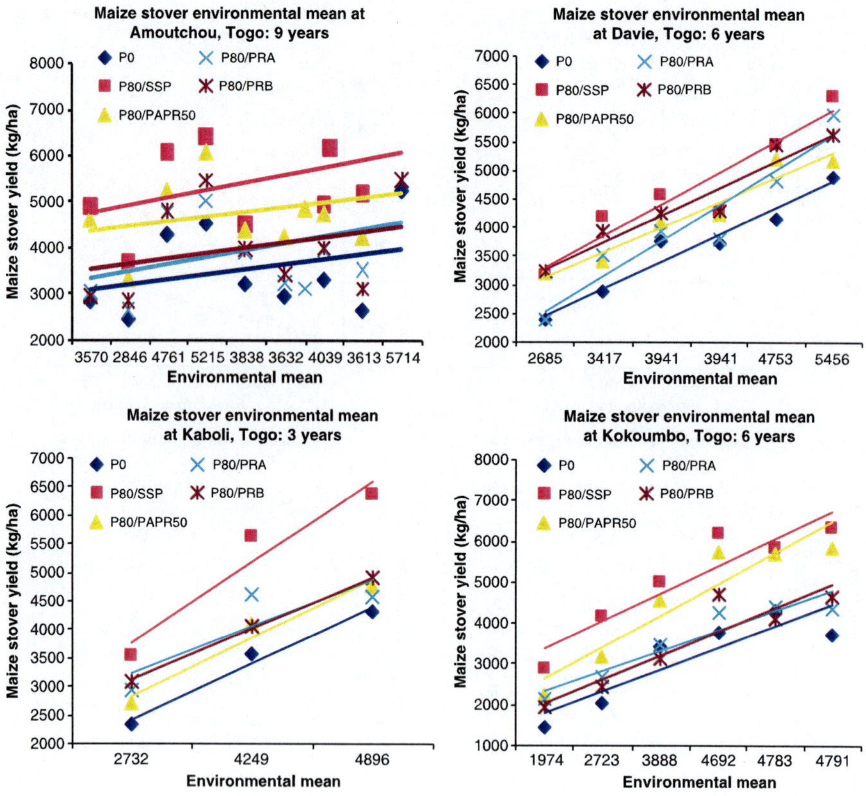

Fig. 9.3 Relationship between Maize stover grain yield and environmental mean in different agro ecological zones of Togo

Conclusion

The present farming systems in West Africa are unsustainable, low in productivity and destructive for the environment. An assessment, initiated by the Food and Agriculture Organization of the United Nations (FAO) in Sub-Saharan countries revealed that the mean annual losses per hectare was 2.5 kg P. As a consequence, the stocks are gradually depleted to the point that they can no longer cope with the demand.

Phosphorus is known to be one of the major constraint to food production in West Africa despite the relative accessibility of these countries to naturally occurring phosphate.

The results of the present paper clearly indicate that for the ago-ecological zones of West Africa the use of PR, PAPR and SSP was profitable as compared to the control treatment. The values obtained for RAE and REI and the most profitable options varied among the experimental sites. It is strongly suggested that future research should put more emphasis on modeling response of crops to the different P sources for the different soil types in the regions.

References

Acquaye DK, Oteng JW (1972) Factors influencing status of phosphorus in surface of Ghana. Ghana J Agric Sci 5:221–228

Jenny F (1973) Experimentation agronomique sur les phosphates naturels de Tilemsi (Mali) en conditions de culture sèche. Agron Trop 28(11):1070–1078

Jones MJ (1973) A review of the use of rock phosphate as fertilizers in francophone, West Africa. Samaru Misc. Paper 43, IARPMB 1044, Zaria, Nigeria

Jones MJ, Wild A (1975) Soil of the West African Savanna. Comm Agric Bureaux Technical Commun No. 55

Milne MK (1968) Annotated bibliographies on soil fertility in West Africa. Common W. Bur Soils, Harpenden

Mokwunye AU (1979) Phosphorus fertilizers in Nigerian savanna soils: II. Evaluation of three phosphate sources applied to maize at Samaru. Trop Agric (Trinidad) 56:65–68

Nabor J, Charoy J, Pichot S (1974) Fertilisation phosphaté des sols du Niger. Utilisation de phosphates naturels de Tahoua. Agron Trop 29(11):1140–1150

Pichot J, Roche AP (1972) Le phosphore dans les sols tropicaux. l'Agronomie Tropicale 27:939–965

Richardson HL (1968) The use of fertilizers. In: Moss RP (ed) The soil resources of tropical Africa. Cambridge University Press, England, pp 137–154

Thibout F, Traore MF, Pieri C, Pichot J (1980) L'utilisation agricole des phosphates naturels de Tilemsi (Mali). Synthèse des résultats de la recherche agronomique sur les cultures vivrières et Oleagineuses. Agron Trop 35(3):240–249

Chapter 10
Long-Term Integrated Soil Fertility Management in South-Western Nigeria: Crop Performance and Impact on the Soil Fertility Status

B. Vanlauwe, G. Nziguheba, O.C. Nwoke, J. Diels, N. Sanginga, and R. Merckx

Abstract Crop response, tree biomass production and changes in soil fertility characteristics were monitored in a long-term (1986–2006) alley-cropping trial in Ibadan, Nigeria. The systems included two alley cropping systems with *Leucaena leucocephala* and *Senna siamea* on the one hand and a control (no-trees) system on the other hand, all cropped annually with a maize- cowpea rotation. All systems had a plus and minus fertilizer treatment. Over the years, the annual biomass return through tree prunings declined steadily, but more drastically for *Leucaena* than for *Senna*. In 2002, the nitrogen contribution from *Leucaena* residues stabilized at about 200 kg N/ha/year, while the corresponding value for *Senna* was about 160 kg N/ha/year. On average, the four *Leucaena* prunings were more equal in biomass as well as in amounts of N, P and cations, while the first *Senna* pruning was always contributing up to 60% of the annual biomass or nutrient return. Maize crop

B. Vanlauwe (✉)
Tropical Soil Biology & Fertility Institute of International Centre for Tropical Agriculture (CIAT-TSBF), ICIPE Duduville Campus off Kasarani Road, P.O. Box 823-00621, Nairobi, Kenya
e-mail: b.vanlauwe@cgiar.org

G. Nziguheba
Tropical Agriculture and Rural Environment, The Earth Institute at Columbia University, 61 Route 9W, Lamont Hall, Room 2G, Palisades, New York 10964,, USA

O.C. Nwoke • N. Sanginga
International Institute of Tropical Agriculture, Ibadan, Nigeria, c/o L.W. Lambourn & Co, 26 Dingwall Road, CR9 3EE, Croydon, UK

J. Diels • R. Merckx
Department of Earth and Environmental Sciences, KU Leuven, Celestijnenlaan 200E - box 2411, B-3001 Leuven, Belgium

A. Bationo et al. (eds.), *Lessons learned from Long-term Soil Fertility Management Experiments in Africa*, DOI 10.1007/978-94-007-2938-4_10,

yields declined steadily in all treatments, but the least so in the *Senna* + fertilizer treatment where in 2002 still 2.2 t/ha of maize were obtained. Nitrogen fertilizer use efficiency was usually higher in the *Senna* treatment compared to the control or the *Leucaena* treatment. Added benefits due to the combined use of fertilizer N and organic matter additions were observed only for the *Senna* treatment and only in the last 6 years. At all other times, they remained absent or were even negative in the *Leucaena* treatments for the first 3 years. Most chemical soil fertility parameters decreased in all the treatments, but less so in the alley cropping systems. The presence of trees had a positive effect on remaining carbon stocks, while they were reduced compared to the 1986 data. Trees had a positive effect on the maintenance of exchangeable cations in the top soil. Exchangeable Ca, Mg and K – and hence ECEC – were only slightly reduced after 16 years of cropping in the tree-based systems, and even increased in the *Senna* treatments. In the control treatments, values for all these parameters reduced to 50% or less of the original values after 20 years. All the above points to the *Senna*-based alley system with fertilizers as the more resilient one. This is reflected in all soil fertility parameters, in added benefits due to the combined use of fertilizer nitrogen and organic residue application and in a more stable maize yield over the years, averaging 2.8 t/ha with maximal deviations from the average not exceeding 21%.

Keywords Alley cropping • Added benefits • *Leucaena leucocephala* • Nitrogen use efficiency • *Senna siamea*

Introduction

A considerable number of long-term experiments have been conducted in Sub-Saharan Africa since the 1920s, and have given valuable insights in soil processes and management practices that control soil fertility (Greenland 1994; Padwick 1983; Pieri 1992; Juo et al.1996).The questions addressed in these trials have shifted with time, and reflect the changes in the research questions arising and the technologies being developed. Models are now increasingly used to evaluate the long-term impact of specific soil fertility replenishment strategies (Parton et al. 1994; Giller et al. 1997), but comprehensive datasets are needed to evaluate the performance of these models. The complexity of agroforestry technologies advocated from the 1980s onwards led to a list of biophysical hypotheses about the functioning and sustainability of these complex systems (Sanchez 1995), some of which hypotheses may only be tested in long-term trials.

Alley cropping was one of the first promising technologies dealing with soil fertility replenishment that was developed and tested at the International Institute of Tropical Agriculture (IITA) during the 1980s and early 1990s. In such cropping systems, food crops are grown between hedges of preferably N_2 fixing trees (Kang et al. 1985). These are regularly cut back to minimize tree-crop competition for water, nutrients, and light. Although it was shown that maize yields in alley cropping systems were better than maize yields in no-tree control treatments, it was acknowledged already at an early stage that addition of chemical fertilizer boosted maize yields further (Kang et al. 1990). Siaw et al. (1991) reported increases in maize production in alley cropping systems after application of N fertilizer, although alley cropping without fertilizer application already doubled maize yields relative to the no-tree control. From alley cropping trials in Benin Republic on Ultisols with low nutrient status in the topsoil, Akonde et al. (1996) concluded that the greatest monetary output was achieved by fertilizer application. Moreover, Schroth et al. (1995) stated that in view of the unbalanced nutrient composition of leguminous mulches in comparison with crop demand, it seems unlikely that fertilizer application can be avoided in continuous agroforestry associations. Notwithstanding the controversy on whether alley cropping is adoptable by the small-scale farmer community, recent surveys have shown that crop-tree intercropping systems have been adopted by a significant proportion of the farmers in Benin, Nigeria, and Cameroon, but in a modified form (Adesina et al. 1999).

In 1986, a long-term alley cropping trial with *Leucaena leucocephala* and *Senna siamea* as hedgerow species was established at the IITA campus in Ibadan to study the long-term crop and tree performance and soil fertility characteristics with and without the input of chemical fertilizer. The cropping system was a maize (first season) – cowpea (second season) rotation. Major conclusions drawn from the first years of the trial (1986–1990) were: (i) *Leucaena* produced most biomass and 20% of the applied N was used by the maize crop, (ii) maize yields were maintained in alley cropping, but the highest yields were obtained in the *Senna* system with fertilizer application, and (iii) the level of the major soil nutrients declined in the no-tree controls, but less so in the alley cropping systems (Van der Meersch et al. 1993).

The current paper describes data on biomass and nutrient accumulation of the hedgerows, maize and cowpea yields, and soil fertility characteristics for the following 16 years (1991–2006). Data for the period 1991–2002 were published earlier by Vanlauwe et al. (2005a). The objectives are (i) to evaluate the long-term biomass production and nutrient accumulation of two different trees (a nitrogen fixing one versus a non N-fixing one) planted as hedgerows and as affected by fertilizer addition, the latter to evaluate their assumed "safety-net" function, (ii) to evaluate the long-term production and stability of crop yields as affected by fertilizer addition and test the hypothesis of improved fertilizer N use efficiency when combined with organic amendments of different quality, and (iii) to assess the impact of long-term application of prunings with or without fertilizer on selected soil fertility characteristics. In an associated report, delta-^{13}C data obtained in this trial were used to evaluate the Rothamsted C model (Diels et al. 2004).

Materials and Methods

Site Description, Experimental Design, and Trial Management

The experiment was conducted at the International Institute of Tropical Agriculture (IITA), Ibadan, Southwestern Nigeria (7°30′N and 3°54′E) on a Ferric Lixisol (FAO 1991) (Table 10.1) with a maximal slope of 5%. A few plots belonging to the lowest replicate are on a Dystric Regosol. Ibadan has a bimodal rainfall pattern

Table 10.1 Selected characteristic of the soil profile at the experimental site in Ibadan, Nigeria (Modified from Vanlauwe et al. 2005a)

	Depth	Org[a] C	Total N	Bray-1 P	pH (H_2O)	pH (KCl)	Exch[a] Ca	Exch Mg	Exch K	Exch Ac[a]	CEC[a]	Sand	Silt	Clay
Soil layer	cm	g kg^{-1}		mg kg^{-1}			$cmol_c$ kg^{-1}					g kg^{-1}		
Ap	0–15	6.2	0.59	4.8	6.2	4.9	1.3	0.3	0.2	0.0	3.0	870	60	70
E	15–40	2.5	0.45	0.7	6.0	4.9	1.5	0.3	0.1	0.6	3.4	750	60	190
EB	40–55	2.8	0.49	0.3	6.1	4.7	2.6	0.5	0.1	0.7	5.6	690	40	270
Bt	55–80	2.5	0.49	0.0	6.1	5.0	3.4	0.7	0.1	0.2	8.5	550	80	370
BC	80–105	3.0	0.50	0.0	6.0	5.2	3.6	0.6	0.1	0.0	8.4	490	100	410
C	105–125	2.3	0.48	0.1	6.4	5.5	4.0	1.1	0.1	0.7	8.0	450	140	410

[a] *'Org'* 'Organic', *'Exch'* 'Exchangeable', *'Ac'* 'Acidity', *'CEC'* 'Cation Exchange Capacity'

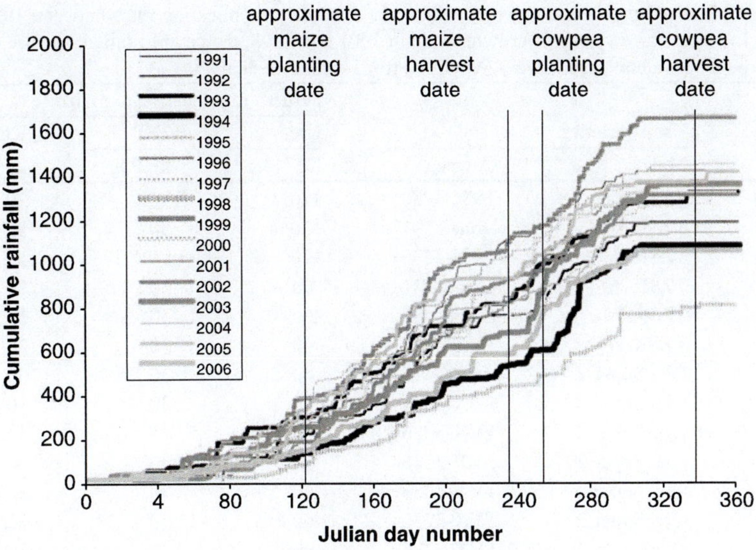

Fig. 10.1 Cumulative rainfall between 1991 and 2006. The vertical bars delineate the approximate maize and cowpea growing seasons (Modified from Vanlauwe et al. 2005a)

(Fig. 10.1). The rainy season starts in March and ends in November, although scanty rainfall occurs until mid-December. The experimental site was established in 1986 as described in detail by Van der Meersch et al. (1993). The trial is a randomized complete block design with six treatments and five replicates, laid out perpendicular to the slope. The treatments are a no-tree control, alley cropping with *Leucaena leucocephala*, and alley cropping with *Senna siamea*, each treated or not with NPK fertilizer. Each alley cropping plot contained five hedgerows and four alleys, 4.5 m wide and 10 m long. *Leucaena leucocephala* is a fast-growing N-fixing legume tree, whereas *Senna siamea* is a non N-fixing legume tree with less vigorous growth characteristics.

Usually, at the start of the first season (April-May) the field was sprayed with herbicide and the hedgerows were pruned a first time. All prunings were applied on the soil surface and the maize was planted at a distance of 0.75 m between the rows and 0.25 m within the rows. Between 1991 and 1996, maize variety TZSR-W was used, while between 1997 and 2006, Oba Super 2 was planted (Table 10.2). At planting, one third of the N fertilizer (urea), and the total amount of P (triple super phosphate) and K (muriate of potash) fertilizer were broadcast over the complete plot. The yearly fertilizer N and P application rates were reduced from 120 to 60 kg N ha^{-1} and from 90 to 30 kg P ha^{-1} in 1994, while K was continuously applied at 30 kg K ha^{-1} (Table 10.2). At about 6 weeks after planting the maize, the hedgerows were pruned a second time. After maize harvest, the field was sprayed with herbicide and the hedgerows were pruned a third time. Cowpea was planted at the same density as the maize and sprayed 3–4 times with insecticides to avoid insect

Table 10.2 Used maize and cowpea varieties and fertilizer application rates between 1991 and 2006. In 1992, no cowpea was grown, while in 1991 and 1998, the cowpea failed because of pest and drought problems respectively (Modified from Vanlauwe et al. 2005a)

	Used varieties		Fertilizer application to maize		
			Urea	TSP	KCl
Year	Maize	Cowpea	Kg N ha^{-1}	kg P ha^{-1}	kg K ha^{-1}
1991	TZSR-W	TVX-3236[a]	120	90	30
1992	TZSR-W	None	120	90	30
1993	TZSR-W	TVX-3236	120	90	30
1994	TZSR-W	IT86D-716	60	30	30
1995	TZSR-W	IT86D-1177	60	30	30
1996	TZSR-W	TVX-3236	60	30	30
1997	Oba Super 2	Ife Brown	60	30	30
1998	Oba Super 2	TVX-3236[a]	60	30	30
1999	Oba Super 2	TVX-3236	60	30	30
2000	Oba Super 2	TVX-3236	60	30	30
2001	Oba Super 2	TVX-3236	60	30	30
2002	Oba Super 2	TVX-3236	60	30	30
2003	Oba Super 2	TVX-3236	60	30	30
2004	Oba Super 2	TVX-3236	60	30	30
2005	Oba Super 2	TVX-3236	60	30	30
2006	Oba Super 2	TVX-3236	60	30	30

[a]Cowpea crop failed because of pests (1991) or drought (1998)

damage to the flowers or pods. The cowpea varieties used each year varied depending on the availability of seeds (Table 10.2). At about 6 weeks after planting the cowpea crop, the hedgerows were pruned a fourth time. Between 1996 and 1999, only the *Leucaena* was pruned a fourth time, as the *Senna* trees did not produce enough biomass to justify a fourth pruning. Usually the field was weeded thrice during the maize season and twice during the cowpea season, of which one weeding takes place immediately before the second and fourth pruning. The number of weedings and prunings and the exact dates of each field activity for each year are presented in Fig. 10.2. Maize and cowpea residues were retained on the plots. In 1991, the cowpea did not yield any beans due to insect damage. In 1992, the field was left fallow for the second season implying that the trees were only pruned twice and that no cowpea crop was grown. Although in 1998 a cowpea crop was planted, it failed because of lack of rain.

Observations

At each pruning activity, except in 2000, the total pruning and wood biomass production per hedgerow was weighed freshly and subsamples of the prunings were taken to determine the dry matter content. At maize harvest, the total ear and stover

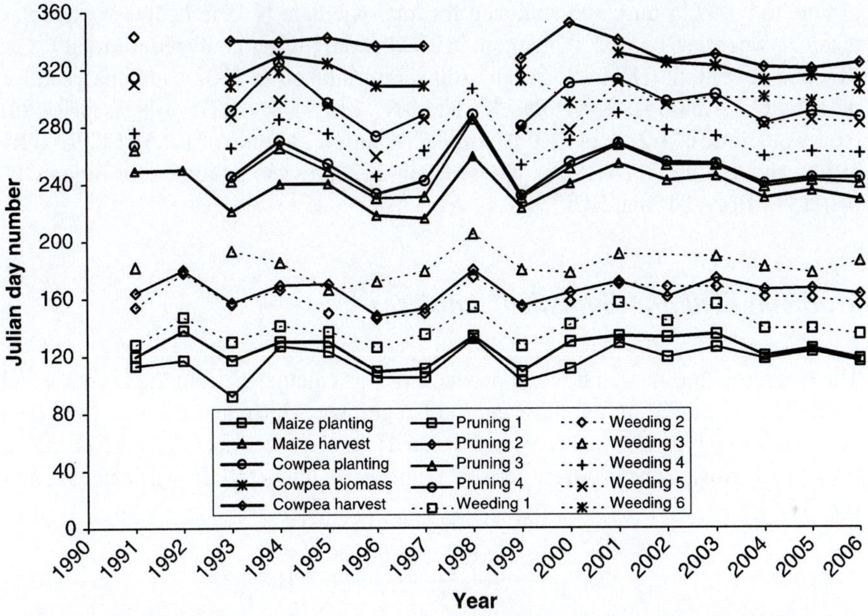

Fig. 10.2 Julian day numbers of the different management activities between 1991 and 2006. In 1992 the minor season was left fallow while in 1998 the cowpea failed to grow due to drought problems. From 1991 until 1995, both hedgerow species were pruned four times yearly (except in 1992, only twice and in 1998, three times). Between 1996 and 1999, the *Senna* trees were only pruned during the first three pruning activities (Modified from Vanlauwe et al. 2005a)

fresh weight was measured on the net plot (borders of 1 m were left on each side of the plots) and a subsample was taken for dry matter determination. In 1992 and 1994, no data on maize stover production are available. As cowpea tends to lose most of its leafy biomass at harvest, quadrants (4.5 by 0.5 m; two quadrants per plot) were used to determine its total aboveground biomass production at physiological maturity. At cowpea harvest, the total pod fresh weight was determined on the net plot and a subsample was taken for dry matter determination. Before the 2002 and the 2006 season, soil samples were taken at 0–5 and 5–15 cm depth, diagonally across the plots (24 cores per plot) and bulked per layer.

Chemical Analyses

The maize, cowpea, and hedgerow samples were air-dried and finally oven-dried (65°C). The maize ear samples were separated into grains and cobs, while the cowpea pod samples were separated into beans and husks. All samples were weighed, ground to pass 0.5 mm, and analyzed for total Kjeldahl N. The hedgerow samples taken between 1991 and 2002 (except in 2000) were further analyzed for total P, Ca, Mg, and K content (IITA 1982). The soil samples taken in 2002 were analyzed for organic C (Amato 1983), total Kjeldahl-N, Bray-1 P (IITA 1982), pH(H_2O) (soil:water ratio of 1:2.5), pH(KCl) (soil:KCl solution 1 M ratio of 1:2.5); ECEC(IITA 1982), and texture (IITA 1982). Those taken in 2006 were analyzed for organic C, pH(H_2O), Bray-1 P, and ECEC.

Mathematical and Statistical Analyses

The hedgerow biomass and wood production was calculated from the biomass and wood data measured on the three central hedgerows. Maize and cowpea yield data in the alley cropping treatments were calculated taking into account the space occupied by the hedgerows. The apparent fertilizer N use efficiency (NUE) was calculated as (Eq. 10.1):

$$NUE = \frac{MaizeN_F - MaizeN_{0F}}{N_{app}} * 100\% \tag{10.1}$$

With $MaizeN_F$ and $MaizeN_{0F}$ the total amount of nitrogen taken up by fertilized and unfertilized maize plants respectively and N_{app} the amount of urea-N applied. Total N uptake comprises N uptake by the grains, cobs, and stover. All units are kg N ha^{-1}. Possible added benefits due to a combination of organic treatments and the mineral fertilizer addition (N_{add}) were calculated as:

$$N_{add} = \left[MaizeN_{prunF} - MaizeN_{prun0F}\right] - \left[MaizeN_{contF} - MaizeN_{cont0F}\right] \ [\text{kgN/ha}] \tag{10.2}$$

With $MaizeN_{prunF}$ the amount of nitrogen taken up by the fertilized, residue (*Senna* or *Leucaena*) treated maize, $MaizeN_{prun0F}$ the amount of N in the corresponding unfertilized stand, $MaizeN_{contF}$ the amount of nitrogen taken up by the fertilized control maize (no prunings) and $MaizeN_{cont0F}$ the unfertilized control. N_{add} so becomes the difference between the extra amount of N taken up in a residue (*Senna* or *Leucaena*) treated, fertilized maize stand and the extra amount of N taken up in the only inorganically treated maize stand (Vanlauwe et al. 2001a). Vanlauwe et al. (2001b) have been pointing to the possible benefits emanating from the combined use of organic and mineral sources of N as an important factor in explaining the improved crop performance sometimes observed under such conditions.

Given the change in N rate and the change in maize variety (Table 10.2), we analysed recovery and added benefits emerging from the combined use of organic and mineral sources of nutrients for three separate categories of data. We grouped together those years where variety TZSR-W was grown with 120 kg N/ha (years 1991–1993) in a first category, the years where the same variety was grown with 60 kg N/ha (years 1994–1996) in a second category and finally the years with variety Oba Super 2 and 60 kg N/ha (years 1997–2002) in a third category.

The MIXED procedure was used to carry out ANOVA on the obtained plant and soil data (SAS Institute 1999). For the data on N use efficiency (NUE, Eq. 10.1) and added benefits (N_{add}, Eq. 10.2), the MIXED procedure was used by category. 'Treatment' was included as fixed variable, while 'Replicate' was included as a random variable. Significant means were separated with the PDIFF option of the LSMEANS statement. Stability analysis (Guertal et al. 1994; Raun et al. 1993) was used to examine the stability of the treatments over time. Treatment means were regressed linearly against annual yields. A mixed model ANOVA was used to test the hypothesis of equality of slopes, using a model similar to one proposed for stability analysis by Stroup et al. (1993). Contrast statements in the ANOVA were used for pair wise comparison of slopes.

Results

Hedgerow Biomass Production, and N, P, and Cation Content

The evolution over the last 16 years of the trial in dry matter pruning biomass and wood yield are given in Fig. 10.3a, b respectively. With respect to pruning biomass production (Fig. 10.3a), the large fluctuations between 1992 and 1993 are due to the fact that in 1992 we only pruned twice, with as a consequence a large increase in the biomass of the first pruning of 1993 and hence in the total biomass yield of that year. This effect is especially obvious for *Leucaena*, less so for the *Senna* stands. The same trend is logically also present in the wood yield (Fig. 10.3b). The graphs showing the pruning-derived addition of N (Fig. 10.4a) or P (Fig. 10.4b) are reflecting this abrupt change very well although more clearly for N than for P. Where the picture for N is almost a copy of the one for pruning biomass added, the picture for P shows a more constant return of P between 7 and 23 kg P/ha, less dependent on biomass yield.

As a general trend, over the years between 1991 and 2006, the amounts of biomass returned (and hence N yield) seem to decline and stabilize at nitrogen additions around 160 kg N/ha for *Senna* and around 200 kg N/ha for *Leucaena*. *Leucaena* maintains the largest N returns, despite the lower biomass yields, compared to *Senna*. Yet, it has to be acknowledged that the overall decrease over the 16 years was larger for *Leucaena* than for *Senna*. The *Leucaena* systems returned on average between 466 and 512 kg N/ha in 1991, while these figures were between 239 and 293 for *Senna* in the same year.

Part of the biomass decline is due to a decreasing number of trees with time. A determination of the number of dead or missing trees in 1999 shows that a large number of the tree rows had more than 25% of the trees missing (Fig. 10.5a). This reduction in the number of trees by 25–50% did not significantly affect the biomass

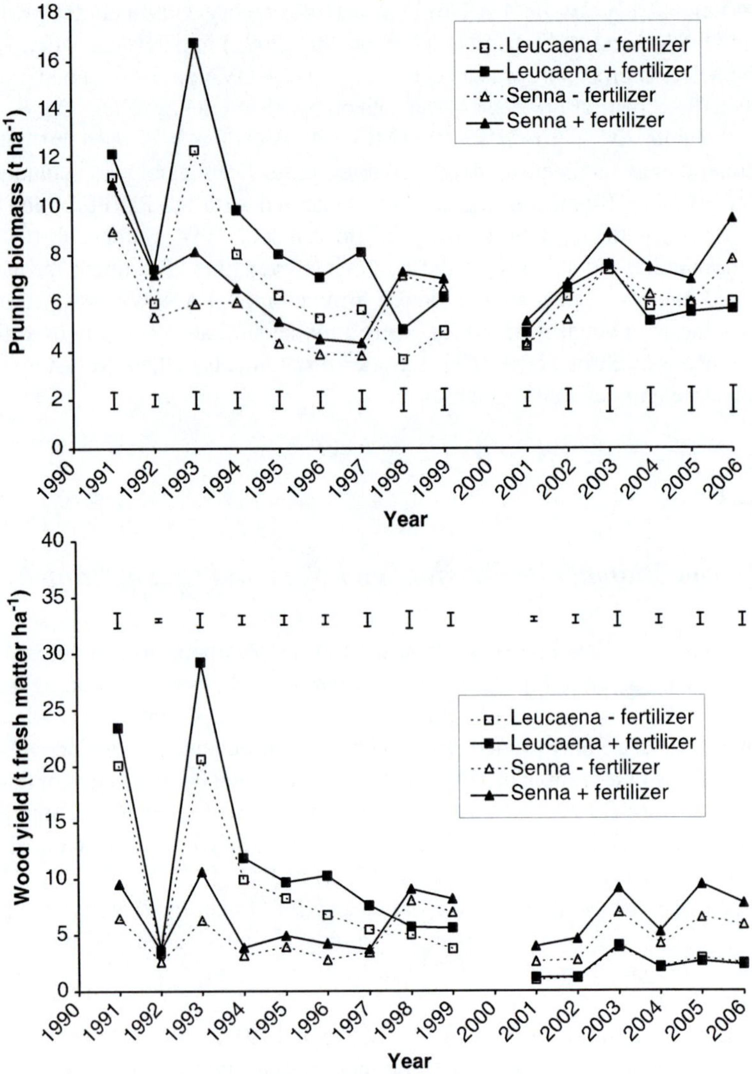

Fig. 10.3 Evolution in pruning biomass added (**a**) and wood yield (**b**), of hedgerow prunings over 16 years between 1991 and 2006, except for missing data in 2000. *Error bars* indicate the SED between treatments for each year (Modified from Vanlauwe et al. 2005a)

production of the Senna rows, while a similar number of missing trees had a drastic effect on the Leucaena rows where biomass already halved (Fig. 10.5b).

With respect to the changes in quality of the prunings within 1 year, a few observations can be made. It can be generalized that the first pruning usually has lower concentrations of N, P and K than the following two or three prunings, which is however

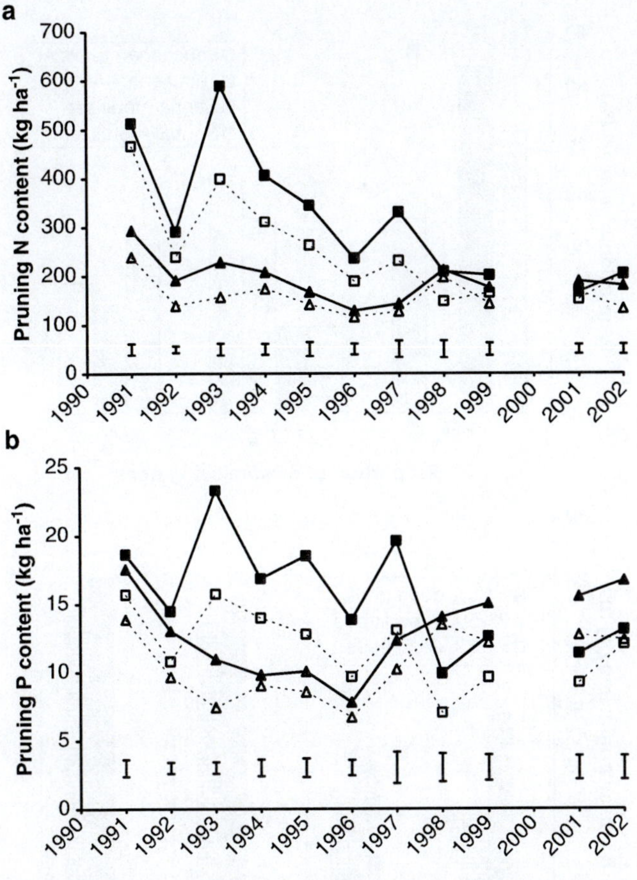

Fig. 10.4 Evolution in nitrogen added through prunings (**a**) and phosphorus added through prunings (**b**) over 12 years between 1991 and 2002, except for missing data in 2000. Error bars indicate the SED between treatments for each year (Modified from Vanlauwe et al. 2005a)

more than compensated by the larger biomass obtained (Tables 10.3 and 10.4). For Ca and Mg, the opposite holds, there the largest concentrations are found in the first pruning of the *Leucaena* stands, while for *Senna* the concentrations of Ca and Mg are rather constant over the four prunings. In Table 10.4, the combination of changes in nutrient contents and biomass yields allows to calculate the proportional contribution of each pruning to either biomass added or to the addition of each specific nutrient. In general, and for all parameters studied, the first pruning contributes most. However, there are extreme differences for specific parameters between the two species. *Senna* is clearly different from *Leucaena* in this that the first pruning always contributes between 50% and 60% of the total amount added for either biomass or any of the nutrients studied. For *Leucaena*, these contributions are much

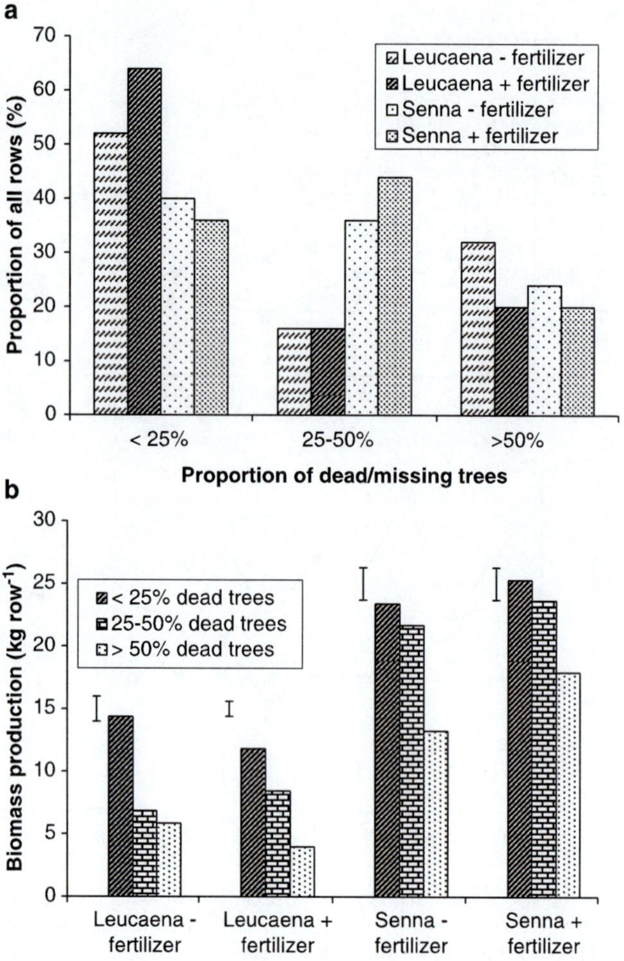

Fig. 10.5 Proportion of dead or missing trees in the two different alley cropping systems as observed in 1999 (**a**) and its effect on biomass production (**b**) (Modified from Vanlauwe et al. 2005a)

more equally distributed over the four pruning events, with only a slightly larger (i.e. 30–35% of the 1st pruning versus 20–25% for the three following ones) contribution from the first pruning.

Selected Soil Fertility Characteristics

In Table 10.5 a selection of several important soil characteristics is given, specifically addressing the effects of the different alley cropping treatments and the combined

Table 10.3 Average (averaged over the years 1991–1998) nutrient concentrations of the four *Leucaena* and *Senna* prunings (Modified from Vanlauwe et al. 2005a)

	N concentration	P concentration	Ca concentration	Mg concentration	K concentration
	% dry matter				
Leucaena – fertilizer					
Pruning 1	3.76	0.178	1.54	0.40	2.14
Pruning 2	4.09	0.215	1.17	0.31	2.58
Pruning 3	3.26	0.191	1.28	0.36	2.39
Pruning 4	4.15	0.218	1.12	0.33	2.39
Leucaena + fertilizer					
Pruning 1	3.74	0.189	1.60	0.41	2.13
Pruning 2	4.14	0.234	1.15	0.30	2.48
Pruning 3	3.72	0.202	1.40	0.37	2.31
Pruning 4	4.24	0.231	1.16	0.33	2.41
Senna – fertilizer					
Pruning 1	2.62	0.179	1.76	0.25	1.41
Pruning 2	3.25	0.250	1.64	0.20	1.95
Pruning 3	3.18	0.218	1.78	0.19	1.78
Pruning 4	3.50	0.184	1.62	0.21	1.92
Senna + fertilizer					
Pruning 1	2.72	0.190	1.76	0.22	1.40
Pruning 2	3.31	0.247	1.62	0.20	1.96
Pruning 3	3.26	0.224	1.79	0.19	1.67
Pruning 4	3.44	0.178	1.66	0.21	1.87
Minimal SED[a]	0.11	0.014	0.08	0.01	0.08
Maximal SED	0.18	0.021	0.11	0.02	0.10

Note that not all trees were pruned and all prunings analysed at all pruning events

[a] '*SED*' 'Standard Error of the Difference'; minimal and maximal SED's are given because each mean comparison test had a different SED depending on the degrees of freedom of both means

Table 10.4 Proportion of the total annual dry matter or nutrient accumulation in the different prunings (Modified from Vanlauwe et al. 2005a)

	Proportion of dry matter or nutrient accumulation			
	Pruning 1	Pruning 2	Pruning 3	Pruning 4
	%			
Total dry matter				
Leucaena – fertilizer	30.7	21.7	25.8	22.6
Leucaena + fertilizer	30.2	21.2	27.1	22.0
Senna – fertilizer	55.6	16.4	17.4	10.9
Senna + fertilizer	59.5	15.0	15.7	10.0
SED[a]	2.0	0.9	1.6	1.0
Total N uptake				
Leucaena – fertilizer	28.5	22.3	25.0	25.2
Leucaena + fertilizer	28.2	21.6	25.5	25.6
Senna – fertilizer	50.0	17.8	19.0	13.6
Senna + fertilizer	54.8	16.5	17.4	11.2
SED	2.3	0.9	1.9	1.3
Total P uptake				
Leucaena – fertilizer	29.4	23.9	22.6	24.4
Leucaena + fertilizer	28.7	23.5	23.0	24.6
Senna – fertilizer	49.6	19.1	17.9	12.6
Senna + fertilizer	55.3	17.8	15.4	10.8
SED	1.3	1.4	1.7	1.1
Total Ca uptake				
Leucaena – fertilizer	35.5	21.2	24.3	20.2
Leucaena + fertilizer	34.7	19.9	28.2	17.4
Senna – fertilizer	54.1	15.8	19.4	11.3
Senna + fertilizer	59.0	13.7	17.4	10.6
SED	2.5	1.5	2.3	1.1
Total Mg uptake				
Leucaena – fertilizer	34.6	18.6	25.5	22.4
Leucaena + fertilizer	35.6	17.5	28.1	19.6
Senna – fertilizer	61.8	14.0	14.3	10.6
Senna + fertilizer	62.2	14.5	13.5	10.2
SED	3.2	1.3	2.0	1.3
Total K uptake				
Leucaena – fertilizer	28.8	24.6	25.9	21.4
Leucaena + fertilizer	28.8	23.2	27.5	20.2
Senna – fertilizer	48.4	20.5	18.1	12.9
Senna + fertilizer	54.5	18.6	15.3	11.8
SED	1.6	1.4	1.4	0.9

Only years where all treatments had four prunings (1991, 1994, and 1995) were included in the analysis. Prunings in 1993 were also excluded as the trees were pruned only twice in 1992

[a]*'SED'* 'Standard Error of the Difference'

effects of fertilizers. The changes in organic carbon and nitrogen content are discussed more extensively in an accompanying paper (Diels et al. 2004) but it can be restated that the organic carbon in all treatments is decreasing between 1986 and 2002 and stabilizing thereafter. Yet, this decrease is less dramatic in the treatments with trees due to the sustained inputs of pruning material over the 20 years. As for the 2002 data, differences between all alley cropping treatments and the respective control treatments with or without fertilizer are highly significant and this holds for both soil layers. The C/N ratios of all treatments seem doubled compared to the values in 1986, due to the considerable loss in soil nitrogen. Between treatments, only in the top 5 cm of the control soil without fertilizer is the C/N ratio significantly higher than any other treatments.

The changes in Bray-P concentrations are rather more reflecting the fertilizer history of the plots than an effect of the trees. Treatments with fertilizer added had, as expected, higher Bray-P concentrations than those without and in both soil layers. Testing for differences between alley treatments with fertilizer and the fertilized control proved significant, but only in the top layer, while the unfertilized alley treatments had similar Bray-P values than the unfertilized controls in both layers. Between 2002 and 2006, Bray-P values continued to decline as the result of reduced P application rates in 1994.

Soil pH reflected the treatments but more convincingly in the deeper layer than in the top layer. Alley treatments resulted in on average about 0.5 pH units higher pH values in the 5–15 cm layers compared to the control soils, both for pH_{H2O} and pH_{KCl}. These differences were not significant in the top soil (0–5 cm) for pH_{H2O} and slightly higher for pH_{KCl} in the alley systems. In both layers the effect of fertilizer additions was significant and resulting in a decrease with 0.2–0.3 pH units as compared to the unfertilized treatments. Cation (Ca, Mg and K) concentrations showed very significant responses to the alley treatments. For both depths and all three cations, alley treatments resulted in larger remaining stocks of these nutrients in soil after 17 years in 2002. This effect is especially obvious for Ca and more specifically in the *Senna* treatments. Testing for differences between the treatments with and those without fertilizers were negative for all the three nutrients and both depths. While there is no significant change in exchangeable acidity with the treatments, the differences in exchangeable cations in the alley treatments lead to higher ECEC values compared to the control soils and for both layers. Differences in texture in 2002 were hardly ever significant.

Maize and Cowpea Grain and Stover Yield and Maize N Uptake

Maize yields have been declining over the years, in an extreme case – the control treatment without fertilizers – even hitting a mere 44 kg/ha in 2002 (Fig. 10.6). Yields were sustained best in the *Senna* + fertilizer treatment with a yield in 2006 of still about 2.5 t of maize per ha. Nevertheless, these remain considerably lower than the initial yield of 3.7 t obtained in 1986 or the peak yield in 1995 of 3.9 t. Yields

Table 10.5 Selected soil characteristics at trial establishment (April 1986 data) and before planting the 2002 maize (April 2002) and the 2006 maize (April 2006) for the 0–5 cm and 5–10 cm (April 1986 data) and 5–15 cm layers (April 2002 and 2006 data) (Modified from Vanlauwe et al. 2005a)

	Organic C		Total N	Bray-1 P		pH (water)		Exch[a] Ca		ECEC[a]		Sand	Silt	Clay
	2002	2006	2002	2002	2006	2002	2006	2002	2006	2002	2006	2002	2002	2002
	$g\ kg^{-1}$			$mg\ kg^{-1}$				$cmol_c \cdot kg^{-1}$				$g \cdot kg^{-1}$		
April 1986														
0–5 cm	12.41		1.45	31.9		5.7		2.8		4.1		872	57	71
5–10 cm	7.51		0.80	28.3		5.2		1.7		2.7		873	48	79
April 2002 or 2006														
0–5 cm														
Control – F[a]	5.7	6.0	0.37	19.5	6.0	5.4	5.6	1.2	0.9	2.2	1.9	812	124	76
Control + F	5.9	6.8	0.48	51.1	28.0	5.3	5.1	1.6	1.0	2.6	2.0	798	130	88
Leucaena – F	9.0	9.5	0.81	13.4	5.3	5.3	5.1	2.1	1.1	3.5	2.3	778	128	84
Leucaena + F	10.0	10.2	0.82	61.0	32.5	5.2	4.9	2.2	1.3	3.7	2.6	770	130	94
Senna – F	10.0	10.2	0.78	18.0	7.4	5.8	6.2	3.4	2.5	4.7	3.7	770	134	70
Senna + F	10.6	13.3	0.84	63.1	32.3	5.4	5.6	3.3	2.7	4.5	3.8	772	124	100
SED[a]	1.1		0.10	6.0		0.1		0.3		0.3		19	11	11
[AC[a] vs control] + F	4.4^{***}		0.35^{***}	10.9^{*}		NS		1.2^{***}		1.5^{***}		NS	NS	NS
[AC vs control] – F	3.9^{***}		0.42^{***}	NS		NS		1.5^{***}		1.9^{***}		-38^{*}	NS	26^{*}
All + F vs – F	NS[a]		NS	41.5^{***}		-0.2^{**}		NS		NS		NS	NS	17^{*}
5–15 cm														
Control – F[a]	3.9	5.0	0.31	13.9	4.4	4.9	4.9	1.0	0.6	1.7	1.6	798	120	94
Control + F	4.2	4.9	0.35	62.3	33.1	4.6	4.5	1.2	0.8	2.0	2.0	770	122	122
Leucaena – F	5.2	6.4	0.44	8.5	3.1	4.9	4.6	1.2	0.7	2.3	1.8	780	118	106
Leucaena + F	5.6	6.1	0.47	62.5	39.6	4.7	4.5	1.3	0.8	2.4	2.0	758	134	108
Senna – F	5.7	7.2	0.43	9.0	3.2	5.9	5.9	2.2	1.8	3.1	2.7	794	112	78
Senna + F	6.7	6.8	0.53	45.8	22.8	5.5	5.5	2.2	2.0	3.1	3.1	770	136	92
SED	0.6		0.05	5.6		0.1		0.1		0.2		16	11	12

[AC vs control] + F	1.9**	0.15**	NS	0.5***	0.6***	0.7***	NS	NS	NS
[AC vs control] – F	1.6**	0.12*	NS	0.5***	0.8***	1.0***	NS	NS	NS
All + F vs – F	NS	NS	46.4***	–0.3***	NS	NS	–24*	14*	NS

`[AC[a] vs control] + F: contrasts tested between all fertilized AC systems versus fertilized control; [AC vs control] – F: idem for unfertilized systems; All + F vs – F: contrasts between all fertilized versus all unfertilized systems

[a] *'Exch'* 'Exchangeable', *'ECEC'* Effective Cation Exchange Capacity', *'F'* 'Fertilizer', *'SED'* 'Standard Error of the Difference', *'NA'* 'Not Available', *'AC'* 'Alley cropping', *'NS'* 'Not Significant'

'*', '**', and '***' = Significant at the 5%, 1%, and 0.1%, respectively

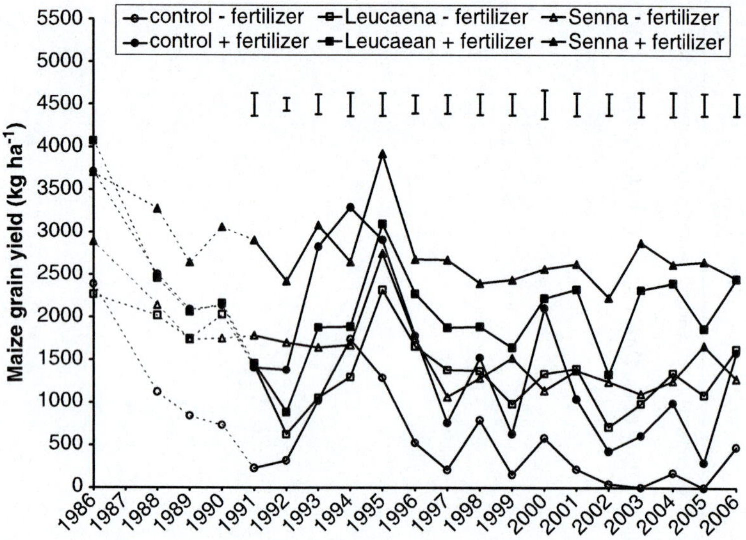

Fig. 10.6 Maize grain yields obtained in the different treatments since the start of the trial in 1986 and up to 2006. Error bars are given from 1992 onwards indicating the SED between treatments for each separate year (Modified from Vanlauwe et al. 2005a)

obtained in the *Leucaena* treatments with or without fertilizer or in the *Senna* treatment without fertilizer varied between 0.7 and 1.3 t per ha (Fig. 10.6).

Stability analysis is a standard tool to evaluate long-term trends in yields (Guertal et al. 1994; Raun et al. 1993). The stability analysis indicated clear differences among the treatments in yield stability over time: the hypothesis of equality of slopes was rejected by the F-test ($P = 0.0007$). A plot of treatment means versus annual production means (Fig. 10.7), revealed that the *Senna* + fertilizer treatment not only sustains the highest average yield as discussed earlier, but also ensures the largest stability (lowest slope, i.e. 0.70). The slope of this treatment was significantly different from the control (maize) with fertilizer ($P = 0.0002$ for H_0: equal slopes) and from *Leucaena* with fertilizer ($P = 0.052$). The slope of the *Senna* + fertilizer treatment was almost equal to and not statistically different from the slopes of the *Leucaena* without fertilizer (0.75) and the *Senna* without fertilizer (0.79) treatments. While those three can thus be termed stable, only the *Senna* + fertilizer also produces an acceptable yield.

Maize nitrogen uptake is given in Fig. 10.8. The data reflect the change in fertilizer addition rate that reduced the N rate from 120 kg N/ha to 60 kg N/ha from 1994 onwards. Between 1991 and 1993, nitrogen uptake increased, remained more or less constant in 1994 and started to decline steadily and levelling out between 1998 and 2006. In general, this mirrors the picture obtained with the maize yields. Largest amounts of N taken up are obtained in the *Senna* + fertilizer system, the lowest ones logically in the control, unfertilized maize.

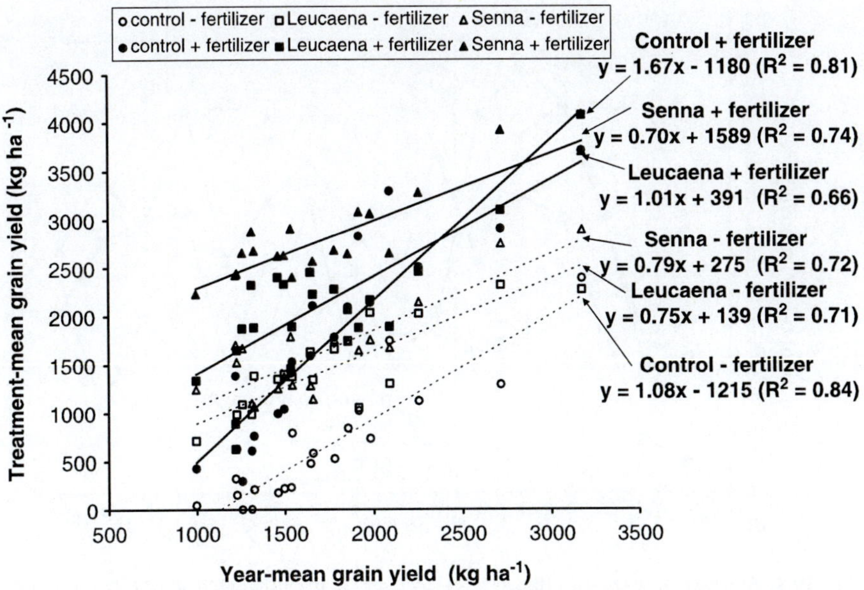

Fig. 10.7 Maize yield stability analysis over the entire duration of the experiment: treatment means versus year means of maize grain yield (Modified from Vanlauwe et al. 2005a)

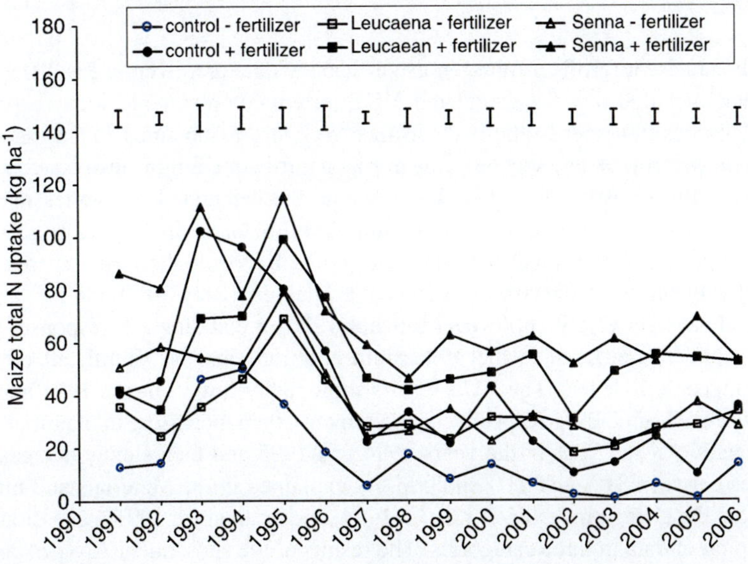

Fig. 10.8 Maize total N uptake in the different treatments between 1991 and 2006 for the different treatments. *Error bars* indicate the SED between treatments for each individual year (Modified from Vanlauwe et al. 2005a)

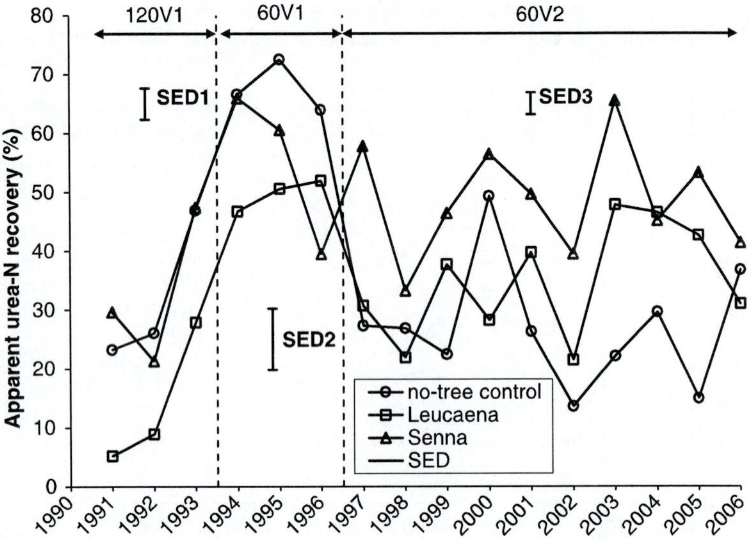

Fig. 10.9 Apparent nitrogen use efficiency (%) by maize in the different treatments between 1991 and 2006 for the different treatments. The error bars indicate the respective SED's between treatments for each category. 120V1 combines years 1991–1993 where maize variety TZSR-W is used with 120 kgN/ha (SED1); 60V1 groups years 1994–1996 where variety TZSR-W is used with 60 kg N/ha (SED2) and 60V2 groups years 1997–2006 where variety Oba Super 2 is grown with 60 kg N/ha (SED3) (Modified from Vanlauwe et al. 2005a)

The apparent fertilizer nitrogen use efficiency data are given in Fig. 10.9. First, we acknowledge that the calculated NUE is termed "apparent" for two reasons. First, it does not refer to the real fertilizer recovery as obtained for instance with labelled N-fertilizer and secondly, the nitrogen fertilizer addition always was accompanied with a dose of P and K. In principle, the calculated recoveries may as a consequence well include a larger amount of soil N taken up by the crops due to a more vigorous crop growth with the extra P and K. While the fertilizer treatments effectively enhanced the concentrations of available P (Bray-I-P, Table 10.5), values were always above critical levels (Sanchez 1976) precluding a P-response. As for potassium, it can be seen that potassium additions had no significant effect on exchangeable K-levels. The NUE results show rather low values in 1991 and 1992 (between 5% and 30% irrespective the treatment), then increasing to maximal values between 46% and 73% in the years 1994 and 1995 and then slowly decreasing to values between 31% and 41% in 2006. As explained in the Materials and methods section, the change in fertilizer rate (in 1994) and cultivar (in 1997) necessitated the grouping of data in three categories. The results of the subsequent ANOVA are presented in Table 10.6. Only for two combinations, 120V1 and 60V2 a significant treatment effect emerged. The first category 120V1, covering the years 1991–1993 showed extremely low NUE for the *Leucaena* systems (14%), and similar NUE

Table 10.6 Apparent NUE and Nadd values for three categories of maize N uptake. Modified from Vanlauwe et al. (2005a)

Treatment	Category		
	120V1[a]	60V1[a]	60V2[a]
	NUE[a] [%]		
Control	32.08	67.68	26.91
Leucaena	14.00	49.74	34.75
Senna	32.81	55.33	48.86
SED[a]	5.30**	10.38^{NS}	3.76***
	Nadd[a] [kgN/ha]		
Leucaena	−21.70	−10.76	4.71
Senna	0.86	−7.41	13.17
SED[a]	5.15***	9.94^{NS}	2.69**

NUE nitrogen use efficiency, cf. Eq. 10.1, *Nadd* nitrogen added benefit, cf Eq. 10.2, *SED* 'Standard Error of the Difference', *NS* not significant
*, ** and *** significant at 0.1%; 1% and 5% respectively
[a]120V1 = Maize variety TZSR-W with 120 kg N/ha (1991–1993); 60V1 = Maize variety TZSR-W with 60 kg N/ha (1994–1996); 60V2 = Maize variety Oba Super 2 with 60 kg N/ha (1997–2006)

values around 32% for the control and *Senna* systems. This adds to the picture that emerged from the total N uptake (Fig. 10.8) which is lower in the *Leucaena* systems as compared with the fertilized *Senna* system. In the 60V1 category, covering the years 1994–1996, no treatment effects on NUE could be found. As from year 1997 onwards, for category 60V2, again significant treatment effects on the NUE values were observed. In line with the better sustained yields in the fertilized *Senna* system, also the highest NUE values were recorded, averaging 49%, compared to 27 or 35% for the control and *Leucaena* treatments respectively.

Analysing for possible added benefits, due to a combination of organic versus mineral sources of N, revealed that the treatments had only significant effects for the first and the last period or category of data, where the data are averaged for the three separate periods (Table 10.6). For 120V1, the first 3 years of the analysis, a negative effect emerged, showing that (i) no added benefit was obtained with *Senna*, while (ii) a negative effect was obtained with *Leucaena*. Introducing *Leucaena* in the system entailed that the increase in N-uptake by maize due to fertilizer addition decreased with 21.70 kg N/ha. Again, in the category 60V1, no significant effects were seen, while in the third period or 60V2 category, (i) no added benefit was found for *Leucaena* (4.71 is not significantly different from 0), while an added benefit of 13.17 kg N/ha was found for the *Senna* systems.

Cowpea grain yields varied between 117 and 1,265 kg ha^{-1} with highest yields occurring in 1995 and the lowest yields in 2001 (*data not shown*). Differences between treatments were usually not significant except in 1995, 1996, and 1999, in which years yields in the *Leucaena* treatments were lowest. The cowpea crop failed in 1991 and 1998 while in 1992 no cowpea was planted. The total aboveground biomass production at peak biomass varied between 241 and 1,536 kg ha^{-1} with highest values in 2002 and lowest in 2001 (*data not shown*).

The underperformance of the N-fixing Leucaena system relative to the Senna system is somehow unexpected. Reasons for this can be found in an enhanced competition between crops and Leucaena trees with consequent inefficient use of N by the crop and ensuing production limitation. The latter has been the subject of a number of earlier studies where the inefficient use of N in a Leucaena based agroforestry systems has been the focus. In (Vanlauwe et al. 1996) we pointed to the low recovery of Leucaena derived N in the maize grains, leaves, stems and cobs totaling 19.4%. Similar data (Vanlauwe et al. 1998b) showed a low recovery of N derived from Leucaena residues of 8.6% in the maize grains, while an effective competition for mineral N between maize and hedgerows was demonstrated in (Vanlauwe et al. 1998a). A second consideration points to the decrease in net primary productivity (trees + crops) in the Leucaena system towards the end of the trial, which can be attributed to a severe tree die-back. Before, the net primary production has been continuously higher in the Leucaena system, up to a few years before the end of the trial, while the primary productivity was only high in the Senna system in presence of N fertilizer.

Discussion

Hedgerow Production

The declining biomass production of the trees in the alley cropping systems provokes questions as to the long-term sustainability of such systems. While the system remains to perform better as far as crop production and soil quality maintenance is concerned, the added value may become too small to make it a rational option. In this study, however, the emphasis was only on monitoring the long-term changes, without providing a complete cost-benefit analysis of the systems involved. One of the reasons of the decline in tree biomass production, apart from the decline in soil fertility, seems to reside in the increasing vulnerability of the trees to the regular pruning as with time, quite large numbers of trees did die (Fig. 10.5a). Again, differences are worth observing between the *Senna* and *Leucaena* stands. While a fall-out of 25–50% of the trees did not significantly affect the biomass production of the *Senna* rows, the same number of missing trees had a considerable effect on the *Leucaena* rows (Fig. 10.5b). Apparently, the *Senna* trees are better competitors to occupy the available above-ground space and close the canopy easily, while *Leucaena* trees fail to do so. The morphology of both trees account for this, *Leucaena* displaying a tall growth pattern, *Senna* having a more widespread canopy. The drastic shifts in biomass production between 1992 and 1993, as stated earlier due to the omission of the two last prunings in 1992, confirms first of all the more even spread of biomass over the four prunings in *Senna*, but may also suggest a better performance and continued growth of *Leucaena* over the dry season 1992–1993.

Soil Characteristics

It can be concluded that perhaps one of the most interesting features – apart from the more documented changes on soil organic C and N – of the alley cropping trials reside in a better conservation of the limited reserves of exchangeable cations like Ca, Mg and K in weathered soils like the one at hand. While this holds for all three nutrients studied here, the effects seem most striking for Ca, more in particular in combination with *Senna*. This has been discussed in another paper (Vanlauwe et al. 2005b) specifically addressing recycling of Ca in *Senna* based alley cropping trials. In view of the absence of a fertilizer effect it must be concluded that the higher concentrations in the two soil layers under alley cropping are due to a reduced loss through leaching and a recycling of the captured nutrients through litter and pruning additions. It has often been argued that trees in cropping systems serve as a kind of safety-net, preventing crop nutrients to leach beyond the rooting zone. The present results provide evidence for such mechanism, but more so for *Senna* than for *Leucaena*. In Vanlauwe et al. (2004), two factors are suggested to explain the difference between the two systems: (i) *Senna* trees recover more Ca from the subsoil than *Leucaena* trees, and (ii) the higher nitrate leaching in the system with N-fixing *Leucaena* trees compared to with the non N-fixing *Senna* trees, as indicated by a long-term N balance, might increase the transport of Ca and other companion cations of nitrate to deeper layers and hence below the rooting zone of the crops.

Grain/Biomass

A general impression that emerges is that the only system that reasonably resists resource degradation while still producing acceptable yields is the Senna + fertilizer system. Soil chemical characteristics, nitrogen use efficiency, added benefits from the combination of organics and fertilizers and yields all confirm this. On top of this, farmers are often more interested in yield stability over the years rather than the occasional peak yield, followed by years with low yields or crop failures. From the stability analysis, it can be concluded that only the *Senna* + fertilizer treatment can be considered to perform at an acceptable level.

Maize N Uptake

The changes in nitrogen fertilizer addition rate in 1994 and the subsequent shift to another maize cultivar in 1997 precluded somehow a straightforward analysis of nitrogen use by maize in the different treatments. However, the re-grouping of the data in sets combining the same cultivar and fertilizer rate reveals an apparent adverse effect of *Leucaena* trees on nitrogen recovery by maize in the first period (120V1).

A recovery as low as 14% is of concern, together with the related observation that combination of an organic and an mineral N source seems to operate in a downward direction. Despite the large supply of N in these systems, the maize crop does not seem to benefit greatly from it, most likely due to competition with the trees. After the transition period (60V1), the superior performance of the *Senna* + fertilizer system emerges, with an average NUE of 49% and an added benefit in the organic/mineral treatments valued at an average of 13.17 kg N/ha. In the same period, none of those effects were obtained with the *Leucaena* system, as both nitrogen use efficiency and added benefits were not significantly different from the control. We suggest that especially the cation recycling characteristics of the *Senna* trees are responsible for the better performance of this system. The evidence of cation recycling, as also found in other trials with *Senna* (Vanlauwe et al. 2004), accounts for the larger NUE values obtained as also mineral nitrogen species can be recycled by the deeper capture of the tree roots. The striking difference between *Senna* and *Leucaena* as regards the proportion the first pruning takes in terms of nutrients added, confirms that *Senna* is able to recover large amounts of nutrients outside the maize growing season and prevents them to leach beyond reach of the maize roots with the following rains.

Conclusions

Comparing the two tree-based cropping systems with *Leucaena* versus *Senna* revealed that, while the former is a nitrogen fixing tree and hence accounts for an extra nitrogen input, the latter provides for a more resilient system. All indicators (soil organic carbon contents and other fertility parameters, maize yields, stability analysis of these yields) underscore this. The main reason for this difference can be ascribed to (i) different rooting patterns of *Senna* versus *Leucaena*, so at the same time being less competitive with maize roots and also recovering nutrients from deeper soil layers (out of reach of maize roots) and/or over the dry period; (ii) to the different qualities of the added pruning material, where slower decomposition seems to be a positive treat, better protecting the soil surface from the impact of rain and temporarily immobilising nutrients so preventing them from leaching beyond the rooting zone and (iii) the tree die-back observed in the *Leucaena* system, reducing primary production towards the end of the investigated period. The data indicate that while nitrogen remains a crucial yield determinant, measures enhancing its use efficiency are often more important than those increasing its inputs. With the present long-term *Senna* trials, it is shown that maintaining all other soil fertility parameters is crucial in this. Maintaining adequate Ca, Mg, K concentrations, preventing pH decline and reducing Al-concentrations all fit under such objective but is a far more difficult target to reach when lime or compound fertilizers are not available or out of reach. Selecting trees for soil fertility maintenance therefore should take those issues into account, rather than focusing on nitrogen inputs only.

Acknowledgements We are particularly grateful to DGDC, the Belgian Directorate General for Development Cooperation, for their sustained support through several and subsequent collaborative projects between K.U.Leuven and IITA on "Soil Organic Matter Dynamics" and "Balanced Nutrient Management Systems for Maize-based Farming systems in the Moist Savanna and Humid Forest Zone of West-Africa", allowing one of the rare and much needed long-term field trials like this. R. Merckx acknowledges support from FWO (Fonds voor Wetenschappelijk Onderzoek, Vlaanderen), IRD (Institut de Recherche pour le Développement) and TSBF-CIAT (The Tropical Soil Biology and Fertility Institute of the "Centro Internacional de Agricultura Tropical") to support part of this work during his sabbatical at IRD/TSBF-CIAT.

References

Adesina AA, Coulibaly D, Manyong VM, Sanginga PC, Mbila D, Chianu J, Kamleu G (1999) Policy shifts and adoption of alley farming in West and central Africa. IITA and Meg Communication, Ibadan, Nigeria

Akonde TP, Leihner DE, Steinmuller N (1996) Alley cropping on an ultisol in subhumid Benin. Part 1: long-term effect on maize, cassava and tree productivity. Agrofor Syst 34:1–12

Amato M (1983) Determination of carbon ^{12}C and ^{14}C in plant and soil. Soil Biol Biochem 15:611–612

Diels J, Vanlauwe B, van der Meersch M-K, Sanginga N, Merckx R (2004) Long-term soil organic carbon dynamics in a subhumid tropical climate: ^{13}C data in mixed C_3/C_4 cropping and modeling with ROTHC. Soil Biol Biochem 36:1739–1750

FAO (1991) World soil resources: an explanatory note on the FAO world soil resources Map at 1:25,000,000 scale. Food and Agriculture Organisation of the United Nations, Rome, 58 p

Giller KE, Cadisch G, Ehaliotis C, Adams E, Sakala WD, Mafongoya PL (1997) Building soil nitrogen capital in Africa. In: Buresh RJ, Sanchez PA, Calhoun F (eds) Replenishing soil fertility in Africa. SSSA, ASA, Waddington, pp 151–192

Greenland DJ (1994) Long-term cropping experiments in developing countries: the need, the history and the future. In: Leigh RA, Johnston AE (eds) Long-term experiments in agricultural and ecological sciences. Proceedings of the conference to celebrate the 150th Anniversary of Rothamsted Experimental Station, Rothamsted, UK, 14–17 July 1993, CAB International, Wallingford, Oxon, UK, pp 187–209

Guertal EA, Raun WR, Westerman RL, Boman RK (1994) Applications of stability analysis for single-site, long-term experiments. Agron J 86:1016–1019

IITA (1982) Automated and semi-automated methods for soil and plant analysis. International Institute of Tropical Agriculture, Ibadan

SAS Institute (1999) SAS OnlineDoc®, Version 8. SAS Institute Inc., Cary, NC, USA

Juo ASR, Franzluebbers K, Dabiri A, Ikhile B (1996) Soil properties and crop performance on a Kaolinitic Alfisol after 15 years of fallow and continuous cultivation. Plant Soil 180:209–217

Kang BT, Grimme H, Lawson TL (1985) Alley cropping sequentially cropped maize and cowpea with leucaena on a sandy soil in Southern Nigeria. Plant Soil 85:267–277

Kang BT, Reynolds L, Atta Krah AN (1990) Alley farming. Adv Agron 43:315–359

Padwick GW (1983) Fifty years of experimental agriculture II. The maintenance of soil fertility in tropical Africa: a review. Exp Agric 19:293–310

Parton WJ, Woomer PL, Martin A (1994) Modelling soil organic matter dynamics and plant productivity in tropical ecosystems. In: Woomer PL, Swift MJ (eds) The biological management of tropical soil fertility. Wiley, Chichester, pp 171–188

Pieri CJMG (1992) Fertility of soils: a future for farming in the West African savannah. Springer, New York, 548 p

Raun WR, Barreto HJ, Westerman RL (1993) Use of stability analysis for long-term soil fertility experiments. Agron J 85:159–167

Sanchez PA (1976) Properties and management of soils in the tropics. Wiley, New York, 618 p
Sanchez PA (1995) Science in agroforestry. Agrofor Syst 30:5–55
Schroth G, Oliver R, Balle P, Gnahoua GM, Kanchanakanti N, Leduc B, Mallet B, Peltier R, Zech W (1995) Alley cropping with Gliricidia sepium on a high base status soil following forest clearing: effects on soil conditions, plant nutrition and crop yields. Agrofor Syst 32:261–276
Siaw D, Kang BT, Okali DUU (1991) Alley cropping with Leucaena leucocephala (Lam.) De Wit and Acioa barteri (Hook. f.) Engl. Agrofor Syst 14:219–231
Stroup WW, Hildebrand PE, Francis CA (1993) Farmer participation for more effective research in sustainable agriculture. In: Ragland J, Lal R (eds) Technologies for sustainable agriculture in the tropics. Proceedings of the two international symposia held in San Antonia, TX and Denver, CO, 1990 and 1991 respectively, ASA Special Publication No.56. American Society of Agronomy (ASA); Madison, Wisconsin; USA, pp 153–186
Van der Meersch MK, Merckx R, Mulongoy K (1993) Evolution of plant biomass and nutrient content in relation to soil fertility changes in two alley cropping systems. In: Mulongoy K, Merckx R (eds) Soil organic matter dynamics and the sustainability of tropical agriculture. Wiley, Chichester, pp 143–154
Vanlauwe B, Swift MJ, Merckx R (1996) Soil litter dynamics and N use in a leucaena (Leucaena leucocephala Lam. (de Witt)) alley cropping system in Southwestern Nigeria. Soil Biol Biochem 28(6):739–749
Vanlauwe B, Diels J, Duchateau L, Sanginga N, Merckx R (1998a) Mineral N dynamics in bare and cropped Leucaena leucocephala and Dactyladenia barteri alley cropping systems after the addition of 15N-labelled leaf residues. Eur J Soil Sci 49:417–425
Vanlauwe B, Sanginga N, Merckx R (1998b) Recovery of Leucaena and Dactyladenia residue nitrogen-15 in alley cropping systems. Soil Sci Soc Am J 62:454–460
Vanlauwe B, Wendt J, Diels J (2001a) Combined application of organic matter and fertilizer. In: Tian G, Ishida F, Keatinge JDH (eds) Sustaining soil fertility in West-Africa. SSSA, ASA, Madison, pp 247–280
Vanlauwe B, Aihou K, Aman S, Iwuafor ENO, Tossah BK, Diels J, Sanginga N, Merckx R, Deckers S (2001b) Maize yield as affected by organic inputs and urea in the West-African moist savanna. Agron J 93:1191–1199
Vanlauwe B, Sanginga N, Giller K, Merckx R (2004) Management of nitrogen fertilizer in maize-based systems in subhumid areas of sub-Saharan Africa. In: Mosier AR, Syers JK, Freney JR (eds) Agriculture and the nitrogen cycle. Assessing the impacts of fertilizer use on food production and the environment. SCOPE 65, ch. 8, Island Press, Washington, DC, pp 115–127
Vanlauwe B, Diels J, Sanginga N, Merckx R (2005a) Long-term integrated soil fertility management in South-western Nigeria: crop performance and impact on the soil fertility status. Plant Soil 273:337–354
Vanlauwe B, Aihou K, Tossah B, Diels J, Sanginga N, Merckx R (2005b) Senna siamea trees recycle Ca from a Ca-rich subsoil and increase the topsoil pH in agroforestry systems in the West African derived savanna zone. Plant Soil 269:285–296

Index

A. Bationo et al. (eds.), *Lessons learned from Long-term Soil Fertility Management Experiments in Africa*, DOI 10.1007/978-94-007-2938-4,